D0867216

Creativity

Creativity

Ethics and Excellence in Science

Mike W. Martin

LEXINGTON BOOKS

A division of
ROWMAN & LITTLEFIELD PUBLISHERS, INC.
Lanham • Boulder • New York • Toronto • Plymouth, UK

LEXINGTON BOOKS

A division of Rowman & Littlefield Publishers, Inc.
A wholly owned subsidiary of The Rowman & Littlefield Publishing Group, Inc.
4501 Forbes Boulevard, Suite 200
Lanham, MD 20706

Estover Road
Plymouth PL6 7PY
United Kingdom

British Library Cataloguing in Publication Information Available

Library of Congress Cataloging-in-Publication Data

Martin, Mike W., 1946–
Creativity : ethics and excellence in science / Mike W. Martin.
p. cm.
Includes bibliographical references and index.
ISBN-13: 978-0-7391-2053-8 (cloth : alk. paper)
ISBN-10: 0-7391-2053-0 (cloth : alk. paper)
1. Research—Moral and ethical aspects. I. Title.

Q180.55.M67M375 2007
174'.95—dc22 2006032240

Printed in the United States of America

∞™ The paper used in this publication meets the minimum requirements of American National Standard for Information Sciences—Permanence of Paper for Printed Library Materials, ANSI/NISO Z39.48-1992.

For Shannon,
and for our daughters,
Sonia *and* Nicole.

Contents

Preface

Creativity has enormous importance in science and technology, yet rarely is it discussed as a topic in the professional ethics of scientists, engineers, physicians, and other science-based professionals. Why is that? Perhaps it is because creativity is assumed to be a non-moral trait which, like intelligence, can be used for good or bad. Or perhaps it is because ethics is assumed to consist solely of general moral requirements incumbent on everyone and on all members of a profession, whereas creativity is assumed to be the gift of a few lucky individuals. In challenging these assumptions, I argue that much scientific and technological creativity constitutes moral creativity, that is, the act or virtue of making new and morally valuable products. Science and science-based professions promote human well-being, improve the environment, enrich our understanding, and augment the meaning scientists find in their work. At the same time, scientific creativity can cause damage in all these areas. For both reasons, creativity deserves to be studied as part of the ethics of science and scientists.

Chapter 1, "Creativity and Ethics," sets forth a broad conception of professional ethics as including ideals, virtues, personal meaning, and good lives, in addition to the responsibilities stated in codes of ethics. Creativity is an ideal rather than a duty, and hence it is neglected when professional ethics is reduced to duties incumbent on all members of a profession.

Chapter 2, "What Is Creativity?," defines creativity as purposefully making new and valuable products. In science these products include significant truths, illuminating explanations, and useful technologies. Scientific creativity qualifies as moral creativity when it contributes to human and environmental goods, whether or not it is motivated by moral aims. There is also a kinship between scientific and artistic creativity.

Chapter 3, "Intellectual Virtues," examines how the virtues contribute to moral creativity. That includes generic moral virtues such as perseverance and self-respect but especially intellectual virtues such as the love of truth, truthfulness, intellectual humility, and intellectual courage. Although creative acts are achievements rather than virtues, the commitment and tendency to be creative becomes a moral virtue when connected with moral aims.

Chapter 4, "Paradoxes of Motivation," reverses direction by emphasizing that desirable outcomes are not always correlated with the virtues. In science as elsewhere, non-moral motives can generate dramatic moral results. This creativity paradox underscores that moral and non-moral goods are interwoven, in science as elsewhere.

Chapter 5, "Serendipity," asks whether the contribution of luck to scientific discoveries alters the moral credit owed to creative scientists. The question connects with broader philosophical concerns about moral luck, that is, luck impacting moral choices. Moral luck does not undermine moral responsibility and instead should quicken our appreciation of how choice and chance interweave.

Chapter 6, "Scientific Misconduct," examines the dark side of creativity. Some scientific misconduct is explicable in terms of greed, arrogance, and other generic human motives. Much wrongdoing, however, arises when the ambition to be creative as a scientist, and to be recognized as such, overshadows respect for truth and responsibilities to colleagues, the public, and one's family.

Chapter 7, "Forbidden Knowledge," asks whether scientists should avoid pursuing truths and topics that seem likely to yield morally undesirable results. Stories such as Prometheus stealing fire from the gods and Adam and Eve eating forbidden fruit caution us that some investigations are too dangerous to pursue, an attitude sharply at odds with the scientific outlook. Can we sustain the scientific ethos of daring inquiry while maintaining ideals of decency and democracy?

Chapter 8, "Leadership," explores morally creative leadership in scientific and technological development. As always, creativity in leadership is identified by reference to valuable outcomes, this time in moving groups toward desirable goals. Also important are morally desirable ways of influencing groups, whether in government projects, university laboratories, or for-profit corporations.

Chapter 9, "Teaching," discusses creativity in education, including innovative approaches to teaching, fostering creativity in students, and improving educational curricula and institutions. Here the valuable new outcomes center on students' mastering and appreciating science, whether in becoming scientists or in acquiring scientific and technological literacy.

Chapter 10, "Good Lives," widens the focus from professional ethics to the ethics of good lives. Good lives are morally desirable, meaningful, and personally satisfying. Ideals of creativity in science contribute to both aspects. These ideals also carry risks, however, and they need to be balanced with other sources of meaning.

The idea for this book began with a course on creativity that I team-taught with chemist Fred Kakis and psychologist Steven Schandler, sponsored by Cameron Sinclair, Chapman's Dean and a professor of religious studies. A more immediate catalyst was an invitation from Henry Shue and Zellman Warhaft to give a paper on professional ethics at Cornell University's College of Engineering. I also wish to thank teachers and colleagues who stimulated my interest in scientific creativity, especially Robert Baum, Nuel Belnap, Jill Buroker, Virginia Carson, Donald Gabard, Brewster Ghiselin, Fred Hagen, Max R. Harward, Fred Kakis, Mark Maier, Sterling McMurrin, Steven Schandler, Roland Schinzinger, and Richard Siddoway.

I thank the editors of journals who granted permission to adapt material from my articles: "Moral Creativity in Science and Engineering, *Science and Engineering Ethics* 12 (2006): 421–33; "Personal Meaning and Ethics in Engineering," *Science and Engineering Ethics* 8 (2002): 545–60; "Paradoxes of Moral Motivation," *Journal of Value Inquiry*, forthcoming; and "Moral Creativity," *International Journal of Applied Philosophy* 20, no. 3 (2006): 55–66.

Above all, I thank my wife Shannon for many conversations about this book—and for her love.

CHAPTER ONE

~

Creativity and Ethics

Scientific creativity is often *moral* creativity. It contributes new and morally valuable products that serve basic human needs for food, energy, and medicine. It promotes understanding of our place in the universe, and it protects nonhuman animals and the environment. Scientific creativity also contributes to the meaning that scientists and other science-based professionals find in their work and lives. At the same time, the technologies it produces generate new moral challenges, from nuclear weapons to global warming, and from computer privacy to human cloning and stem cell research. In addition, creative ambitions sometimes draw individuals and institutions into wrongdoing. For all these reasons, creativity deserves to be part of the study of ethics in science and science-based professions.

Usually it is not. Although creativity is implicitly understood to be part of research ethics, rarely is it taken up as a moral topic in its own right.[1] The moral aspects of creativity draw us into neglected issues in professional ethics, including moral luck, forbidden knowledge, paradoxes of moral motivation, and creative teaching and leadership. It also sheds new light on standard topics such as scientific misconduct and responsibilities to the public. In exploring these issues we need a broad conception of professional ethics that includes not only the duties incumbent on all members of a profession, but also personal commitments and ideals, especially the ideal of being creative.

Creativity in science takes many forms. It includes discovering significant truths, developing illuminating explanations, designing useful technologies, teaching science in innovative ways, and showing leadership in research and development. In this book I understand science broadly to include the natural and social sciences, technology (both hardware and knowhow), scientific and technological organizations, and science-based professions like engineering and medicine. Of course, in other contexts we distinguish "pure science," aimed at understanding, from "applied science," aimed at technology. Today they are thoroughly interwoven, however, and certainly in thinking about creativity we need to consider their connections. The results of basic science constitute "scientific capital" that can be used long into the future in stimulating technological development.[2] And technological goals make possible the massive funding required for large-scale scientific inquiries, in addition to sometimes inspiring advances in understanding the world.

Personal Meaning and Public Goods

As members of the public, we tend to think of science and technology in terms of their dramatic and ubiquitous practical impacts. Consider the National Academy of Engineering's list of the preeminent engineering achievements during the twentieth century: electrification, automobiles, airplanes, water supply and distribution, electronics, radio and television, agricultural mechanization, computers, telephones, air-conditioning and refrigeration, highways, spacecrafts, the Internet, imaging technologies, petrochemical technologies, laser and fiber optics, nuclear technologies, and high-performance materials.[3] In appreciating the public impacts of science and technology, however, we should not neglect the moral importance of scientific understanding itself and the personal commitments of scientists, engineers, and other science-oriented professionals.

Sometimes the creative contributions of scientists spring directly from their moral commitments. Gertrude B. Elion, for example, focused her career on cancer research.[4] This commitment began with her choice of chemistry as a college major, a choice she made in direct response to the slow and painful death of her beloved grandfather from stomach cancer. As she recalls, "that was the turning point. It was as though the signal was there: 'This is the disease you're going to have to work against.' I never really stopped to think about anything else."[5] Like other women during the 1930s and 1940s, she faced tremendous obstacles, not the least of which was being denied entrance to doctoral programs despite having graduated from college with highest honors.

Elion sought work that provided opportunities to do research. When she finally found a job at a drug research company which accepted women because of the shortage of male workers during World War II, she became a passionate and tireless investigator. The drugs she discovered saved thousands of children with leukemia, in some cases advancing them from certain death to an 80 percent survival rate. She received a Nobel Prize for helping to develop new research techniques for comparing nucleic acid metabolism in normal and abnormal cells, techniques that greatly accelerated the development of new drugs. Throughout her writings she expresses joy in helping, and gratitude for being able to help—attitudes that sustained her passion for science. She also inspired a generation of women to pursue careers in science.[6] One of her colleagues suggests that she did more good for the human race than Mother Teresa.[7]

Jonas Salk's moral contributions provide a more familiar illustration.[8] A chemistry course during his first year of college inspired him to become a doctor, not in order to heal sick people but instead to find ways to prevent them from getting sick in the first place. In medical school he developed a special interest in vaccines, and he took a year's leave to pursue an opportunity to do research. After medical school he helped develop the first effective vaccine for influenza, which was killing more American soldiers than combat. Later, after becoming head of his own laboratory, he began to study polio. At the time, polio killed and maimed tens of thousands of children each year. Parents lived in terror of the disease, and during the yearly epidemic parents would drastically curtail their children's activities in order to protect them. The disease struck adults as well, most famously President Franklin Delano Roosevelt. In discovering a vaccine for polio, Salk pioneered the general possibility of vaccinations that have more long-lasting effects than traditional immunizations. All these endeavors were infused with a sense of responsibility to use science to reduce suffering.

Some scientists risk their lives in pursuing creative endeavors. That is true of Marie Curie and Pierre Curie, whose research involved daily handling of radioactive material long before its dangers were known. It is true of physicians who have been willing to become the first experimental subjects in testing new drugs that potentially have dangerous side effects. And it is true of Peter Palchinsky, who illustrates creative leadership in science and engineering.[9]

During the early twentieth century, Palchinsky crusaded for the rights of workers and the safety of the public affected by technology. Educated as a mining engineer, his first job was studying workers in a coal-mining operation in the Ukraine. He discerned that the efficiency and productivity of the mines was linked to the workers' living conditions, and he developed the first quantitative information about their poor housing and transportation. The experience transformed him. Throughout his three-decade career, he continued to understand technical matters in engineering as interwoven with social, economic, and political issues. Gradually moving into top leadership positions, he lobbied for engineers to become more broadly educated and to accept wider responsibilities for the human dimensions of their work. Although his only crime was zealousness in humanitarian engineering, Stalin had him executed for treason.

Even when their work is not a direct response to human suffering, most scientists and engineers are influenced in their creative endeavors by moral commitments. Some professionals, for example, turn away from work they find morally objectionable, such as weapons development. Gene Moriarty reports that his first job offer after college was in a large aerospace company.

> The engineers in my prospective group were excitedly telling me about a system they were developing. It sensed the terrain with an ingenious radar mechanism, employed an elaborate feedback control structure, and made determinations on the basis of statistical decision rules. The job offered fascinating prospects for sophisticated engineering designs. But then I took a wider look at the project and realized that the system I'd be working on was to form part of the signal processing unit of what came to be the Cruise Missile.[10]

Moriarty decided not to pursue the job because he believed that the only good of war was to make rich people richer.[11] In contrast, engineers with commitments to a strong national defense in order to safeguard democratic values would have responded quite differently, especially if they saw in the cruise missile an accurate weapon that could minimize civilian casualties. Personal commitments and ideals, then, should not be dismissed as mere subjective matters lacking relevance to professional ethics. They enter centrally into how individuals understand their responsibilities to the public affected by their work.

Moral commitments are not the primary impetus behind most creative activities, however. Far more important, even more important than money, is the excitement of the work itself. As an example, consider the development of Data General Corporation's Eclipse MV/8000 computer, as recounted in Tracy Kidder's *Soul of a New Machine*. The innovative computer was developed in record time. It was made possible by a group of thirty engineers who were excited by the project and handpicked by the project director. Kidder portrays their motives as akin to those of stonemasons who built cathedrals during the Middle Ages as temples to God. Like them, the engineers were engaged in "the sort of work that gave meaning to life" by evoking enthusiasm and bringing a sense of fulfillment.[12] As a contrast, Kidder ends his book by quoting a regional sales manager who, in giving a pep talk to the sales staff marketing the computer, asks rhetorically, "What motivates people?" The manager answers his own question: "Ego and the money to buy things that they and their families want."[13] Ego and money motivated the engineers as well, but their primary motivation was the excitement of technological development, with the additional stimulus of camaraderie. This case reminds us that professions involve more than doing competent work in providing useful products and services. Professions also foster creative expression of character—and soul, in Kidder's sense of vitality and spirit. As the example also illustrates, even non-moral motives and commitments have moral relevance in stimulating commitments to excellence.

Creative individuals value creativity. They organize and structure their lives around endeavors where they hope to make new and valuable contributions. As Santiago Ramon y Cajal writes, they have "a strong inclination toward originality, a taste for research, and a desire to experience the incomparable gratification associated with the act of discovery itself."[14] D. N. Perkins adds, "people who *want* to be creative, who deeply value such a characteristic in themselves, are more likely to make themselves creative and keep themselves that way."[15] In doing so, however, they depend on supportive colleagues and contexts in which their creative aspirations can flourish. Although I often use individuals as examples, it should be borne in mind that contemporary science is generally conducted by groups of individuals. Thus, although Nobel Prizes for a discovery are restricted to at most three people, they typically recognize the chief scientists heading laboratories in which dozens and even hundreds of individuals contributed to the discovery.[16]

Creativity and Professional Ethics

The preceding examples barely hint at the myriad ways in which commitments and ideals of creativity enter into professional ethics.[17] Why, then, is creativity rarely included in the study of professional ethics in science? One reason concerns stereotypes of creativity as anarchic freedom, or a mysterious and indefinable process, which cannot be contained within the regimen of professionalism.[18] I challenge these stereotypes in chapter 2, where I understand creativity in terms of purposeful endeavors leading to new and valuable products. Another reason is that creativity is assumed to be a non-moral trait having no special connection with morality. I reject that view in chapter 3 by understanding the intellectual excellences contributing to creativity as themselves moral virtues.

Still another source of neglect of creative and other personal commitments concerns stereotypes of professional ethics. Professionalism is associated with setting aside personal values in order to sustain professional objectivity. Of course, professional distance requires rigorous control of biases that could undermine objectivity. Yet, the passion for objectivity and the reasoned devotion to professional standards are themselves personal commitments of great importance in science and technology. Moreover, personal commitments dramatically shape careers and contribute to professionalism through career choices, job choices, project choices, and the sheer ambition to be creative.

Another stereotype of professional ethics reduces it to the duties set forth in professional codes of ethics. According to this "consensus paradigm," professional ethics consists entirely of a profession's consensus about the moral requirements mandatory for its members, together with the study of the foundations of those requirements and the ethical dilemmas generated when they conflict with each other.[19] These codified requirements have great importance in ensuring uniform standards of responsibility among members of a profession. Nevertheless, exclusive attention to them eclipses the moral significance of creativity, which by its nature cannot be a duty incumbent on everyone. More generally, the emphasis on codified duties in thinking about professional ethics neglects character, virtues, ideals, and personal commitments.[20]

Personal commitments and ideals include humanitarian, environmental, religious, political, aesthetic, supererogatory, and family commitments. They also include voluntary commitments to professional standards, especially when these commitments are linked to an individual's broader value perspective. As Bernard Williams writes, personal commitments enter into "a set of desires, concerns, or . . . projects, which help to constitute a *character*."[21] These commitments and their accompanying ideals motivate, guide, and give meaning to the work of scientists and other professionals, in addition to contributing to public goods for the wider community.

Personal *meaning* connotes both intelligibility (making sense) and value. These dual connotations are connected, for we make sense of our lives in terms of the values we live by. In this way, meaning connects directly with motivation and commitment, with what we care deeply about in ways that sustain our interest and energy, shape our identities, and create pride or shame. Ultimately these commitments and meaning are linked to living good lives, lives that are morally desirable and personally satisfying, in science as elsewhere, as I discuss throughout this book and most directly in chapter 10.

Meaning has both subjective aspects (a sense of meaning) and objective aspects (genuine value and coherence). A sense of meaning is a practical disposition or attitude manifested in emotions and motivation. Creative individuals experience hope for progress, frustration at delays, joy in discovery, delight in confirmation of results, pride in achievement, curiosity about how the simple and the complex interact, bafflement about anomalies, admiration (as well as envy) for others' achievements, disgust and contempt for shoddy work. These "rational passions" are as crucial to creative work as technical skills.[22] In addition, a subjective sense of meaning can be assessed as enlightened or distorted in light of the objectively justified values that make work worthwhile and life worth living. And individual meaning is closely linked with institutional involvements, as Joanne B. Ciulla reminds us.

> Meaningful work, like a meaningful life, is morally worthy work undertaken in a morally worthy organization. Work has meaning *because* there is some good in it. . . . Work makes life better if it helps others; alleviates suffering; eliminates difficult, dangerous, or tedious toil, makes someone healthier and happier; or aesthetically or intellectually enriches people and improves the environment in which we live.[23]

Using this broad conception of professional ethics as including personal commitments and meaning, we can now discuss more fully what creativity is and how it is linked to moral creativity.

Notes

1. For example, see Kristin Shrader-Frechette, *Ethics of Scientific Research* (Lanham, MD: Rowman & Littlefield, 1994); and Edmund G. Seebauer and Robert L. Barry, *Fundamentals of Ethics for Scientists and Engineers* (New York: Oxford University Press, 2001).
2. Vannevar Bush, *Science—The Endless Frontier* (Washington, D.C.: NSF, 40th anniversary ed., 1990), 19.
3. National Academy of Engineering, www.greatachievements.org. Discussed by Mike W. Martin and Roland Schinzinger, *Ethics in Engineering*, 4th ed. (Boston: McGraw-Hill, 2005), 7.
4. Sharon Bertsch McGrayne, *Nobel Prize Women in Science*, 2nd ed. (Washington, DC: Joseph Henry Press, 1998), 279–302.
5. McGrayne, *Nobel Prize Women in Science*, 286.
6. Gertrude B. Elion, "Personal Reflections," in *Women in Science and Engineering: Choices for Success*, ed. Cecily Cannan Selby (New York: New York Academy of Sciences, 1999), 16–18.
7. McGrayne, *Nobel Prize Women in Science*, 296.

8. J. S. Smith, *Patenting the Sun: Polio and the Salk Vaccine* (New York: William Morrow and Company, 1990), 149, 282–89; and Jeffrey Kluger, *Splendid Solution: Jonas Salk and the Conquest of Polio* (New York: Berkley Books, 2004).

9. Loren R. Graham, *The Ghost of the Executed Engineer*, (Cambridge, MA: Harvard University Press, 1993).

10. Gene Moriarty, "Ethics, *Ethos* and the Professions: Some Lessons from Engineering," *Professional Ethics* 4, (1995): 77.

11. Gene Moriarty, "Ethics, *Ethos* and the Professions," 77.

12. Tracy Kidder, *The Soul of a New Machine* (New York: Avon, 1981), 273.

13. Kidder, *The Soul of a New Machine*, 291.

14. Santiago Ramon y Cajal, *Advice for a Young Investigator*, trans. Neely Swanson and Larry W. Swanson (Cambridge, MA: MIT Press, 1999), 48.

15. David N. Perkins, *The Mind's Best Work* (Cambridge, MA: Harvard University Press, 1981), 271.

16. R. Keith Sawyer, *Explaining Creativity: The Science of Human Innovation* (New York: Oxford University Press, 2006), 270.

17. For additional examples see Caroline Whitbeck, *Ethics in Engineering Practice and Research* (New York: Cambridge University Press, 1998), 306–12; Stephen Monsma, C. Christians, E. R. Dykema, L. Arie, S. Egbert, and V. P. Lambert, *Responsible Technology: A Christian Perspective* (Grand Rapids, MI: William B. Eerdmans Publishing, 1986); and Mary Tiles, & Hans Oberdiek, *Living in a Technological Culture* (London: Routledge, 1995), 172–75.

18. For example, David B. Resnik, *The Ethics of Science* (New York: Routledge, 1998), 38; and Paul K. Feyerabend, *Against Method: Outline of Anarchistic Theory of Knowledge* (Atlantic Highlands, NJ: Humanities Press, 1975). On creativity as an indefinable process, see David Bohm, *On Creativity* (New York: Routledge, 1996), 1.

19. I challenge the consensus paradigm in *Meaningful Work: Rethinking Professional Ethics* (New York: Oxford University Press, 2000).

20. That neglect began to change with groundbreaking books such as Alasdair MacIntyre, *After Virtue*, 2d ed. (Notre Dame, IN: University of Notre Dame Press, 1984); and Edmund L. Pincoffs, *Quandaries and Virtues* (Lawrence: University Press of Kansas, 1986.)

21. Bernard Williams, "Persons, Character and Morality," in Bernard Williams, *Moral Luck* (New York: Cambridge University Press, 1981), 5.

22. Williams, "Persons, Character and Morality," 118.

23. Joanne B. Ciulla, *The Working Life: The Promise and Betrayal of Modern Work* (New York: Times Books, 2000), 225–26.

CHAPTER TWO

What Is Creativity?

To be creative is to discover or invent new and valuable products. In science the products include significant truths, explanations, problem-solving techniques, and technologies (which include not just hardware but also practical know-how and organizational structures). Accordingly, what counts as creativity turns on judgments about what is new and what is valuable, and those judgments are fallible and contestable. Scientific creativity often constitutes moral creativity, and it has parallels with morally creative decision making in general, as well as with creativity in the arts.

New and Valuable Products

Early psychological studies portrayed creativity in terms of special personality traits and mental processes, sometimes called "lateral thinking," that abandon straight-ahead logical reasoning.[1] Most recent studies, however, reject that view. Creativity involves quite ordinary processes and traits, including mastery of relevant skills, curiosity, willingness to experiment, open-mindedness, imagination, flexibility, ability to draw analogies, tolerance of uncertainty, perseverance, and ambition. Furthermore, a genius IQ is not required for creativity in science; usually the IQ of a typical college graduate (about 120) suffices.[2] Conversely, many exceptionally intelligent people are not especially creative.

Instead of unusual processes and personalities, creativity needs to be understood in terms of outcomes or *products*.[3] In turn, creative *processes* and *activities* are those which lead (or tend to lead) to discovering and inventing creative products; creative *persons* and *groups* invent or discover creative products; creative *organizations* develop creative products; creative *working conditions* are conducive to creative products; and so on. What, then, are creative products? They are new and valuable outcomes which emerge from purposeful activities.

New means original, as assessed relative to some appropriate point of reference. Usually the reference point is a domain, that is, an area of inquiry and activity with a knowledge and skill base that needs to be mastered in order for creativity to take place.[4] Thus, creative biology consists in discoveries that advance the domain of biology, and creative engineering consists of innovation in electrical, chemical, mechanical, or some other branch of engineering. Even when discoveries have broader import, generic domains can be specified; for example, the invention of the wheel advanced transportation, and the discovery of ways to make fire advanced energy technology. Occasionally creativity involves establishing an entirely new domain—for example, establishing computer science as a discipline during the twentieth century.

Within a given domain, alternative reference points can be used to identify what is new. In *history-relative* creativity, which I usually have in mind, a product is new relative to preceding accomplishments in human history.[5] In *person-relative* creativity, a product is new relative to an individual's previous accomplishments or level of development—an idea especially relevant to teaching creativity. And in *group-relative* creativity, a product is new relative to a specified group or society. Thus, Euclid manifested history-relative creativity because he was the first person to develop the geometry bearing his name. An exceptionally talented youth would be person-relative creative if she reinvented parts of Euclidian geometry without having encountered it before; she would also be group-relative creative because she makes discoveries beyond those of most members of her peer group.

When two thinkers make the same discovery within close periods of time, they might both be credited with being history-relative creative, as were Charles Darwin and Alfred Russel Wallace for discovering evolution, Newton and Leibniz for discovering calculus, and Jack Kilby and Robert Noyce for co-inventing the microchip. More often, however, only the first discoverer or inventor is recognized as history-relative creative. As we will see, this emphasis on being first has considerable importance for the ethics of scientific creativity.

Valuable means important, innovative, interesting, useful, or otherwise desirable in terms of the values within a domain. In theoretical science the central values are epistemic (truth-centered), for example, significant truth, explanatory power, predictive power, simplicity, and beauty. In applied science and engineering the key values include usefulness, safety, health, efficiency, aesthetics, and legality.

Purposeful activities means that intentional endeavors are involved in some direct way, allowing that luck also plays a key role. Typically some form of ingenuity or insight is employed. The proverbial chimpanzee who accidentally types a novella after millions of random keystrokes on a computer is not being creative. What about computers that generate pictures, music, writing, and math having some merit?[6] Is the computer creative, or only the person who programmed it? I would say only the person is. But whichever way we answer, we would be unlikely to use such examples as paradigms in teaching the concept of creativity to a child. Even in clear-cut cases, however, the purposeful activity should be understood broadly. In "found art," for example, the intention is simply to place an everyday object on display for aesthetic appreciation or engagement. The intentions in science and engineering are usually clear enough, for example, to explain a puzzling occurrence, solve a mathematical problem, or develop a new source of energy.

Products include artifacts, ideas, techniques, and organizations. In theoretical science the most important products are explanatory theories or models that are empirically testable and have predictive power. In applied science and engineering the important products are safe and useful technological products and processes. As noted in chapter 1, increasingly theoretical science and technological development are interwoven. For example, theoretical science cannot be advanced without precise scientific instruments, and in turn theoretical science contributes to engineers' understanding of materials and energy. Some creative advances, such as the invention of transistors, are creative advancements in both science and engineering.

The term "products" needs to be understood broadly to include creating new problems, as well as creating solutions to problems. Frequently the most important creative step, the one that most quickly leads to fresh ideas, consists in providing a clear statement of the relevant problem.[7] Alternatively, the problem might be sufficiently clear, and the creative move consists in appreciating its importance. Given limited resources and time, decisions about which investigations to pursue can make all the difference.

Criticisms and Clarifications

The definition of creativity, as making or discovering new and valuable products, is entirely formal. It specifies a thin concept, a mere skeleton of creativity, and it does so by listing the features of central cases of creativity, rather than with a set of precise necessary and sufficient conditions. As such, it leaves open several questions concerning assessments of history-relative creativity.

To begin with, who has the authority to decide what is valuable and new? Although each domain has its field of experts, disagreements can arise concerning who qualifies as an expert or a leading authority. Such disagreements are more pronounced in the arts and humanities, where there is less agreement about standards of excellence, but they are not absent from science. Even when a consensus exists about who the experts are, there can be disagreements about the judgment of the experts in a given case. Experts are human, and sometimes biases distort their judgment, for example, by inflating their assessments of creativity in order to obtain research funding for themselves or friends. More subtle distortions arise when the thinking of all experts in a field are dominated by powerful assumptions or paradigms, as Thomas Kuhn calls them. Even leading experts sometimes make glaring mistakes, as in the failure to appreciate the work of Emily Dickenson, Vincent van Gogh, and Gregor Mendel until long after their deaths.[8] And especially in technological development, what is celebrated as a creative contribution might later appear far less valuable. Perhaps an example is the invention of gas-guzzling, high-polluting, and dangerous sport utility vehicles (SUVs).

Mihaly Csikszentmihalyi would have us go further in understanding creativity as radically relative to the pronouncements of acknowledged authorities. According to his "systems model," creativity is inseparable from its recognition: "Creativity results from the interaction of a system composed of three elements: a culture that contains symbolic rules, a person who brings novelty into the symbolic domain, and a field of experts who recognize and validate the innovation. All three are necessary for a creative idea, product, or discovery to take place."[9] Csikszentmihalyi implies that Dickinson, van Gogh, and Mendel became creative only after they died, when their work was recognized and validated by a field of experts.[10] Indeed, should they ever again fall out of favor with the experts they would then cease to be creative.

Csikszentmihalyi deserves credit for underscoring the importance of a field of experts in judging creativity within a domain, as well as to the historical and cultural context in understanding how creativity arises. But making creativity wholly relative to experts is implausible. Surely van Gogh and Mendel were creative while they were alive! It is just that their contributions were not appreciated until after they died.

The values and standards used in judging creativity might be challenged as unwarranted and subjective. The most fundamental controversy centers on skepticism about the possibility of objectively defensible values, and hence about the reliability of assessments of creativity based on those values. Radical skeptics reject the existence of an objective, external world that we can reliably come to know. They renounce the notion there is any one reality, and they contend that human understanding is always relative to conceptual schemas, none of which is more valid than others. These skeptics include the postmodernists who provoked the "science wars" of recent decades.

If "subjective" means irremediably biased, we have a quick reply to the skeptics. There is nothing biased or subjective about the truth that DNA is composed of nucleic acids in a double-helix structure, or that human beings evolved from lower life forms. Nor is there anything biased about the moral value of developing new drugs to cure deadly human diseases, improving public health, and developing new sources of energy. A bias is a distortion of truth and truth-seeking, and hence the very notion implies the possibility of sometimes getting at the truth. Epistemic and moral values can be justified, as can the ideas of truth, objective reality, and accurate explanation, thereby providing a basis for assessing scientific creativity. Of course, this quick reply will not satisfy the skeptics, and much subtle philosophy is devoted to providing a fuller response.[11] Here I simply note the obvious: Our view of values, truth, and understanding shapes our assessments of creativity.

A softer skepticism, one more sympathetic to the aspirations of science, does not renounce the possibility of objective truth, but instead is agnostic about that possibility. Thomas Kuhn is such a skeptic. Famously, he distinguishes between normal and revolutionary science. Normal science takes place within a dominant paradigm, where "paradigm" means either (1) a dominant exemplary achievement within a domain of science, such as Newton's laws or Einstein's theory of relativity in physics, or (2) a larger set of ideas, methods, and models that guide work in a discipline.[12] Revolutionary science, in contrast, consists in overthrowing and replacing a dominant paradigm, as in the shift from Newton's laws to Einstein's relativity to quantum mechanics. Kuhn tries to make sense of both normal and revolutionary science in terms of consensus and controversy within scientific communities, without relying on the ideas of objectivity and truth. His distinction between normal and revolutionary science can also be used to distinguish two types of scientific creativity: (a) "puzzle solving" within a dominant paradigm, and (b) revolutionary overthrow of a dominant paradigm and replacement of it with a new paradigm that literally "changes the world" we live in.

Despite his powerful insights, Kuhn has been subjected to a barrage of criticisms.[13] His claim that normal science is dominated by one dominant paradigm is frequently untrue, for usually several influential paradigms compete within a domain at a given time. Moreover, there is far greater continuity across paradigm changes than Kuhn recognizes. And his heady talk of paradigm changes "creating a new world," as distinct from new experiences of the world, has been criticized as romantic hyperbole. For my purposes, however, the main criticism is that Kuhn's agnosticism cannot make sense of how paradigms are ultimately accepted or rejected because of how they fit the facts. Ideas such as useful explanations, accurate predictions, and unifying perspectives all presuppose that science succeeds in uncovering truths about an objective reality. Intellectual humility implies accepting our fallibility in the pursuit of truth, but it presupposes there is truth to be humble about.

Some skeptical critiques of science do contain elements of truth—truth of the sort they claim to do without. Skeptics force us to appreciate that science is not the value-neutral enterprise it was once assumed to be. Moral, religious, political, and even aesthetic values influence the direction of scientific investigation. Especially today, science is a collective activity requiring massive funding, and funding decisions typically take into account value perspectives about which investigations are most promising in terms of meeting human needs and gaining new understanding. Those value perspectives might be more or less enlightened and fair. Disagreements have arisen, for example, concerning the relative funding of research on gender-related diseases, such as ovarian and breast cancer versus testicular and prostate cancer. We can acknowledge a host of values that guide science, so long as we include (significant) truth among those values. As Philip Kitcher urges, we need an integrated perspective that is sensitive to the influence of values on science while affirming the "science realism" claim that "the sciences sometimes deliver the truth about a world independent of human cognition, and they inform us about the constituents of that world that are remote from human observation."[14]

Finally, and this is not so much a criticism as an amplification, some innovations might fall into several domains where they are judged according to different sets of values. Thus, sociobiology and evolutionary psychology are relevant to virtually all human sciences and humanities, but in different ways and with different degrees of importance. Similarly, a work of art might be judged as part of the domain of sculpture, but also as part of the domain of economically valued objects. Its aesthetic value might be minimal, but if it is the work of Picasso it will have considerable financial value. And a technique or technology for breaking open safes might be valuable in the domain of criminal activities but not in the domain of marketing safes.

Some scholars deny that creativity is involved in the example of safecrackers—ingenuity perhaps, but not creativity. A. J. Cropley, for example, believes that moral permissibility is a necessary condition for creativity: It is "repugnant to speak of the creativity of a cheat, a mass murderer, or an evil demagogue."[15] In reply, we can grant there is an air of irony in calling ingenious thieves, torturers, and terrorists creative, for we wish to avoid any hint of praising them. We can take account of Cropley's point, however, by noting that creativity takes many forms, some more valuable than others. Creativity does not always mean moral creativity.[16]

Scientific Creativity as Moral Creativity

Moral creativity consists in discovering or making new and *morally* valuable products. For example, in everyday life, morally creative outcomes include new and valuable ways to appreciate morally relevant facts, integrate conflicting values, identify priorities among conflicting moral values in response to ethical dilemmas, educate or inspire others to appreciate the importance of moral issues, and develop new institutions and laws. Extending earlier terminology, we can think of moral creativity as generating new and valuable products in the *moral domain*, recognizing that the moral domain is interwoven with all areas of life. As responsible moral agents, each of us can be person-relative and group-relative creative in bringing morally valuable goods into existence. History-relative moral creativity comes as a direct response to moral challenges within groups or societies, especially in situations that are morally uncertain, indeterminate, and ambiguous, and where moral precedents are either lacking or contested.

Scientific creativity is also moral creativity when it provides new and morally valuable products. Such products include the deepened understanding of the world, useful technological goods, and social progress. Sometimes the products come in direct response to human suffering, as with Gertrude Elion's development of new drugs and Jonas Salk's development of a polio vaccine. Improvement in the quality of life is illustrated by twentieth-century engineering marvels, such as automobiles, jets, and computers. And Peter Palchinsky illustrates morally creative leadership that contributes to social progress.

In addition, consider what Alvin M. Weinberg calls "Quick Technological Fixes," or quick fixes to social problems.[17] Social problems typically arise when many individuals behave in undesirable ways. Changing their behavior directly might be impossible or unfeasible without sacrificing other important values, in particular freedom. Technology can alleviate the problem. For example, when people drive their cars too fast, the cost of increasing law enforcement might be prohibitive. Creatively engineering safer roads and safer cars, with seat belts and air bags, proves more effective in reducing accidents and injuries. Again, poverty is due in part to unjust distributions of wealth, but it might be more socially acceptable to creatively engineer less expensive, genetically modified foods than to raise taxes and redistribute wealth.

Weinberg noted that quick fixes, like all technology, typically generate new problems. Even then, however, new and better science and technology can remedy flawed technology. A dramatic illustration is the research done by Sherwood Rowland and Mario Molina, who in 1995 received the first Nobel Prize (in chemistry) for work done in environmental science. Chlorofluorocarbons (CFCs) are human-made chemicals widely used since the 1940s in aerosol cans and refrigeration (as Freon).[18] During the 1970s Rowland and Molina showed that CFC molecules exist for decades and rise to the ozone layer that surrounds earth and protects it from the sun's ultraviolet rays. There the CFCs chemically interact to break down ozone, in ways described in a beautiful set of equations published by the two chemists in 1974. Alarmed by the implications of their research, Rowland and Molina spearheaded a public relations campaign that challenged a $2-billion-a-year industry led by DuPont Corporation. By 1978 the Environmental Protection Agency set forth regulations for ending the manufacture of CFCs. The ban quickly stimulated the development of more environmentally friendly technologies replacing CFCs. The problem has not disappeared, and harm caused by existing CFCs will continue for decades, but the ozone layer is beginning to repair itself.

Even when scientific creativity is not the key to solving moral problems, it provides information relevant to creative moral reasoning. Consider recommendations for public policy regarding stem cell research. In 1978 the first test-tube baby was born in England. The process involves fertilizing several eggs, typically leaving spare embryos at the end of the process. What should be done with these extra embryos? In particular, is it permissible to experiment on them in order to gain knowledge about the safety of in vitro fertilization and other important matters (including, as we now know, stem cell research)? Science cannot answer these questions, but it can provide information relevant to personal decisions and social policy.

In 1982 England established a committee charged with advising British ministers on possible legislation concerning cloning intended for therapeutic purposes, as distinct from reproductive purposes. Mary Warnock, a distinguished philosopher, was put in charge of the committee.[19] Her first act was to have medical experts inform the committee about the science involved. This helped the committee appreciate that they were confronting a genuinely new situation which creative science had created but could not resolve. Never before in human history had a fertilized human egg been viable outside a woman. The issue was not when life begins, for all cells are living and (except for sperm and ova) possess a full set of DNA. The issue was to find a morally acceptable social policy for how to treat unused embryos.

The committee recommended allowing experimentation on unused embryos, with the consent of their owners, for fourteen days only, at which time they would be destroyed. Several reasons supported this recommendation. On the one hand, the committee decided to allow some experimentation, given how promising the research was, and given that the embryos would eventually be destroyed in any case. On the other hand, the committee sought a firm time limit in order to allay the public's fear of a slippery slope in the direction of experimenting on more advanced fetuses. The fourteen-day limit was somewhat arbitrary, although it coincided with the time embryos usually embed in the uterine wall and begin accelerated development. Yet, the limit was also based on their view that the embryo is not sentient. Most members of the committee also became convinced that it is not yet a person with rights to life. Obviously this last belief is controversial. Liberals on abortion regard the committee's work as a creative compromise; conservatives on abortion will refuse to see the committee as morally creative at all. (Hopefully, a new quick fix is emerging that provides new stem cell lines without destroying embryos.)

Models of Moral Creativity

The Warnock committee illustrates several aspects of creative moral decision making about public policies in science (and elsewhere). They include pinpointing key issues, making relevant scientific inquiries, responding to others' views, making reasonable compromises, imaginatively assessing options, and exercising good judgment in interpreting and integrating relevant moral values. Of course, much routine moral decision making also involves these activities. Moral creativity occurs when the activities lead to valuably new decisions and results.

Moral creativity does not mean inventing moral values from scratch.[20] That muddled notion was promulgated by Jean-Paul Sartre, who contended that "moral choice is comparable to the construction of a work of art."[21] Sartre writes, "my freedom is the unique foundation of values and . . . *nothing*, absolutely nothing, justifies me in adopting this or that particular value, this or that particular scale of values."[22] This view, that there is no right or wrong choice or moral perspective, beyond the demand for authenticity in acknowledging our freedom to choose, makes nonsense of morality. If we follow Sartre in characterizing moral decision making as inventing, we must at least restore the ideal he abandons—*responsible* inventing, inventing that *reasonably* takes responsibilities (obligations) into account.

Sartre's art analogy is less helpful, however, than an analogy with scientific creativity. Creative discoveries in science arrive at insights into the way the world is. When Watson and Crick identified the structure of DNA, they got it right: They discovered, not invented, the truth. Likewise, most moral decision making involves making discoveries, although this time the creative product is a morally acceptable course of action rather than a scientific truth.

Creativity in engineering provides an even better metaphor and model for moral creativity. As Caroline Whitbeck points out, engineers begin with goals that might be more or less clear, in situations that are more or less indeterminate.[23] Sometimes their most important task is to define or redefine the problem they face.[24] They must accept and integrate multiple design constraints, including available materials, cost and profit, usefulness to clients, and legal and moral constraints. They take into account general scientific knowledge and previous solutions to similar problems (benchmarking). Sometimes there is one right course of action, and moral creativity consists in discerning it. Other times there are several acceptable results. Because engineers work in groups, they must routinely take into account others' viewpoints and remain open to reasonable compromises with colleagues, employers, and clients. All these features of the engineering model apply to the Warnock committee. The committee was faced with the task of integrating multiple moral ideals, ranging from respect for individuals' rights to protection of public health, and they proceeded by reaching reasonable compromises and consensus based on input from many participants.

American pragmatists frequently apply engineering metaphors to understand moral creativity and moral reasoning. In particular, John Dewey invoked metaphors of production and design, construction and reconstruction, tools and "experimental engineering," to characterize intelligent and creative moral reasoning.[25] Rather than applying fixed rules in a cookbook fashion, a reasonable moral decision "coordinates, organizes and functions each factor of the situation which gave rise to conflict, suspense and deliberation."[26] More recently, Jeffrey Stout characterizes creative moral reasoning as "moral bricolage" in which we use whatever moral and conceptual resources we have ready at hand, "taking apart, putting together, reordering, weighting, weeding out, and filling in" so as to arrive at reasonable solutions to problems.[27] James Wallace adds that moral decision making requires "intelligent, calculated improvisation and the virtue of resourceful inventiveness in adapting our practical knowledge to unprecedented difficulties."[28]

Although pragmatism accurately portrays Warnock's creative decisions, it does not provide a full account of the values involved. In my view, pragmatism complements but does not replace other moral traditions, each of which has enriched our moral understanding. We benefit from a plurality of moral traditions in creativity shaping our future.

Beauty in Art and Science

Creativity in science differs from creativity in the arts, so much so that some scholars restrict the term "creativity" to the arts and use "discovery" for the sciences. Creativity in the arts typically involves expressing and conveying emotions through music, painting, sculpture, novels, film, or other artifacts. In contrast, discovery in the sciences consists in finding impersonal truths that, once discovered, are communicated in journals that rarely mention the emotions motivating scientists. According to Arthur I. Miller, "Artists bare their hearts and souls to the world in a product that is intensely personal, while modern scientists are constrained to hide their hopes, dreams, aspirations, and angst in their personal correspondence and unpublished manuscripts."[29]

These stereotypes are only partly accurate, however, and they do not justify restricting the concept of creativity to the arts. Creativity, defined as generating new and valuable products, applies equally to the sciences and the arts. Furthermore, elements of discovery, making, and invention can enter into creativity in all domains. Thus, a scientist might invent a fresh model to explain a range of phenomena, and invent a new symbolism to express the model mathematically. Again, classification schemas in botany and zoology combine discovery with invention. Conversely, artists discover new patterns and communicate truth, including universal truths about human experience.[30] The truths tend to be about human experiences, but even science must connect with emotionally resonant human experiences and observations, whether directly or indirectly.

Furthermore, the creative products of science have aesthetic dimensions. Once outside technical journals, scientists frequently comment on the beauty of their discoveries. They express astonishment, awe, delight, excitement, love, pleasure, serenity, surprise, wonder, terror, reverence, and a sense of mystery. Beauty is not reducible to these emotions. Instead, beauty consists of aesthetic features of the world and scientific representations of the world which are valued as worthy of being cherished. These features include order and harmony, unity and wholeness, form and pattern, simplicity and elegance, symmetry and proportion, splendor and sublimity, vastness and variety. In this way, beauty has both subjective and objective aspects, and engages both emotions and aesthetic appreciation, whether we are talking of science or the arts.

Francis Hutcheson distinguished three categories or levels of abstraction of scientific beauty.[31] The first level consists of living creatures and natural objects and forces. Appreciating the beauty of a galaxy or the structure of the brain is akin to appreciating a work of art. The second level is comprised of regularities in nature, as set forth in theories and models that describe patterns and explain regularities. The third level is beauty in mathematical formulations, both those of pure mathematics and those used to formalize scientific theories.

Mathematical simplicity, which condenses complexity and chaos into elegant formulas such as $e=mc^2$ and $f=ma$, is the most familiar paradigm of beauty in science. Simplicity can mean different things, however.[32] Economy of symbolism might mean (1) using few symbols, as in the examples from Newton or Einstein, or instead it might mean (2) using the fewest symbols necessary to express the truth. Formulas in quantum mechanics and string theory are enormously complex, and hence not simple in the first sense, but they might be simple in the second sense. Again, simplicity might refer to structural features of the world, including perceived grace and flow, rather than to mathematical representations.

Moreover, as in the arts, albeit on a lesser scale, scientists disagree in their aesthetic judgments and sensibilities. In *Beauty and Revolution in Science*, James W. McAllister notes that in mathematics the dominant sensibility is a classical one of "unity, economy, symmetry, consistency, balance, harmony, order, and the life," but in biology and history there is a more Romantic "aesthetic that values diversity, differentiation, complexity, and organicism."[33] Even Hutcheson, in developing his theory of beauty as uniformity amidst variety, emphasized that the focus of beauty might be either uniformity or variety: "The figures which excite in us the ideas of beauty seem to be those in which there is *uniformity amidst variety*. . . . What we call beautiful in objects, to speak in the mathematical style, seems to be in compound ratio of uniformity and variety: so that where the uniformity of bodies is equal, the beauty is as the variety; and where the variety is equal, the beauty is as the uniformity."[34]

In science, simplicity is subordinate to truth. Here we should distinguish between the pursuit and confirmation of truth. The search for simplicity, and more generally for beauty, partly motivates the pursuit of truth of many researchers, and it is also one factor in choosing among competing theories. Paul Dirac famously said "it is more important to have beauty in one's equations than to have them fit experiment."[35] That is hyperbole. Many theories strike scientists as beautiful but are later overthrown as false. Einstein revered mathematical beauty and worked with a sense of nature as mysterious, but he emphasized that "experience remains, of course, the sole criterion of the physical usefulness of a mathematical construction."[36]

As McAllister emphasizes, scientists should and generally do hold paramount the shared standards in assessing the truth of scientific theories: empirical content and testability, internal coherence, consistency with empirical data, power to predict new data, consistency with other well-corroborated theories, and explanatory power. Aesthetic concerns should only influence scientific theories by complementing and underscoring truth-centered criteria. McAllister shows that aesthetic criteria are themselves applied in the degree that theories are promising as explanations of empirical truths and as compatible with other corroborated theories. That is, what counts as a beautiful explanation depends in part on how well a theory is perceived to fit the facts. A similar relationship between simplicity and truth holds in disciplines that apply science, such as architecture and engineering: "The empirical success of a theory contributes to determine the weighting of that theory's aesthetic properties within the community's aesthetic canons."[37] The notion of ubiquitous ether through which electromagnetic radiation flows smoothly seemed beautiful *until* it was shown to be contrary to experimental evidence.

In light of this preeminence of truth and understanding in science, we can now turn more fully to the truth-centered virtues and ideals that guide scientists. Doing so will help us see why these virtues and ideals not only contribute to moral creativity but are themselves moral in nature.

Notes

1. Robert W. Weisberg, *Creativity: Genius and Other Myths* (New York: W.H. Freeman and Company, 1986), 51–69; R. Keith Sawyer, *Explaining Creativity: The Science of Human Innovation* (New York: Oxford University Press, 2006), 39–74.

2. Dean Keith Simonton, *Creativity in Science* (New York: Cambridge University Press, 2004), 103–104.

3. Vincent Tomas, "Introduction," in *Creativity in the Arts*, ed. Vincent Tomas, (Englewood Cliffs, NJ: Prentice-Hall, 1964), 1–3; Simonton, *Creativity in Science*, 15.

4. J. Baer, "Domains of Creativity," in *Encyclopedia of Creativity*, ed. Mark A. Runco and Steven R. Pritzker (San Diego: Academic Press, 1999), 591–96.

5. Margaret A. Boden, *The Creative Mind*, 2nd ed. (New York: Routledge, 2004), 2.

6. R. Keith Sawyer, *Explaining Creativity: The Science of Human Innovation*, 97–111.

7. H. Scott Fogler and Steven E. LeBlanc, *Strategies for Creative Problem Solving* (Upper Saddle River, NJ: Prentice Hall, 1995), 1; and Mihaly Csikszentmihalyi and

R. Keith Sawyer, "Creative Insight: The Social Dimension of a Solitary Moment," in *The Nature of Insight*, ed. R. J. Sternberg and J. E. Davidson (Cambridge, MA: MIT Press, 1995), 329–63.

8. The details of the Gregor Mendel case are contested. See J. Waller, *Einstein's Luck: The Truth Behind Some of the Greatest Scientific Discoveries* (New York: Oxford University Press, 2002), 132–58.

9. Mihaly Csikszentmihalyi, *Creativity* (New York: Harper Collins, 1996), 6, 29–30.

10. Cf. Nancy C. Andreasen, *The Creating Brain* (New York: Dana Press, 2005), 14–16.

11. For example, see *The Nature of Truth*, ed. Michael P. Lynch (Cambridge, MA: MIT Press, 2001).

12. Thomas S. Kuhn, *The Structure of Scientific Revolutions*, 3rd ed. (Chicago: University of Chicago Press, 1996), 175.

13. See Alexander Bird, *Thomas Kuhn* (Princeton, NJ: Princeton University Press, 2000); and Peter Godfrey-Smith, *Theory and Reality* (Chicago: University of Chicago Press, 2003), 75–101.

14. Philip Kitcher, *Science, Truth, and Democracy* (New York: Oxford University Press, 2001), 28.

15. A. J. Cropley, *More Ways than One: Fostering Creativity* (Norwood, NJ: Albex, 1992), 49.

16. Cf. Carl Rogers, "Toward a Theory of Creativity," in *The Creativity Question*, ed. Albert Rothenberg and Carl R. Hausman, (Durham, NC: Duke University Press, 1976), 297; and Raymond S. Nickerson, "Enhancing Creativity," in *Handbook of Creativity*, ed. Robert J. Sternberg, (New York: Cambridge University Press, 1999), 396–397; and Robert Nozick, *The Examined Life* (New York: Simon and Schuster, 1989), 35.

17. Alvin M. Weinberg, "Can Technology Replace Social Engineering?" *University of Chicago Magazine* 59 (October 1966): 6–10.

18. Sherwood Rowland, "Ozone Hole," in *Life Stories*, ed. Heather Newbold, (Berkeley: University of California Press, 2000), 134–40; and Aisling Irwin, "An Environmental Fairy Tale: The Molina-Rowland Chemical Equations and the CFC Problem," in *It Must Be Beautiful: Great Equations of Modern Science*, ed. Graham Farmelo, (London: Granta Books, 2003), 87–109.

19. Mary Warnock, *Nature and Morality: Recollections of a Philosopher in Public Life* (New York: Continuum, 2003), 69–110.

20. Cf. Mary Midgley, "Creation and Originality," in Mary Midgley, *Heart and Mind: The Varieties of Moral Experience* (New York: St. Martin's Press, 1981), 54.

21. Jean-Paul Sartre, "Existentialism Is a Humanism," trans. P. Mairet in *Existentialism from Dostoevsky to Sartre*, ed. Walter Kaufmann, (New York: New American Library, 1975), 364.

22. Jean-Paul Sartre, *Being and Nothingness*, trans. Hazel E. Barnes (New York: Washington Square Press, 1966), 76.

23. Caroline Whitbeck, *Ethics in Engineering Practice and Research* (New York: Cambridge University Press, 1998), 53–73.

24. Fogler and LeBlanc, *Strategies for Creative Problem Solving*, 1–5.

25. John Dewey, *Human Nature and Conduct* (New York: The Modern Library, 1957 [1922]), 139. Cf. Larry A. Hickman, *John Dewey's Pragmatic Technology* (Bloomington: Indiana University Press, 1990), 111.

26. John Dewey, *Human Nature and Conduct*, 183.

27. Jeffrey Stout, *Ethics After Babel: The Languages of Morals and Their Discontents*, rev. ed. (Princeton, NJ: Princeton University Press, 2001), 74. Stout borrows the bricoleur model from Claude Levi-Strauss, who contrasts the bricoleur (Jack of all trades) with the engineer (a professionalized designer), a contrast Stout downplays. Claude Levi-Strauss, *The Savage Mind* (Chicago: University of Chicago Press, 1966), 16–17.

28. James D. Wallace, *Moral Relevance and Moral Conflict* (Ithaca, NY: Cornell University Press, 1988), 93–94.

29. Arthur I. Miller, *Insights of Genius: Imagery and Creativity in Science and Art* (Cambridge, MA: MIT Press, 2000), 430.

30. Gunther S. Stent, "Meaning in Art and Science," in *The Origins of Creativity*, ed. Karl H. Pfenninger and Valerie R. Shubik, (New York: Oxford University Press, 2001), 31–42.

31. Francis Hutcheson, *An Inquiry Concerning Beauty, Order, Harmony, Design*, ed. Peter Kivy (The Hague: Martinus Nijhoff, 1973). Discussed by James W. McAllister, *Beauty and Revolution in Science* (Ithaca, NY: Cornell University Press, 1996), 18–21.

32. Gideon Engler, "Aesthetics in Science and in Art," *British Journal of Aesthetics* 30, no. 1 (January 1990): 29–30.

33. James W. McAllister, "Introduction to Recent Work on Aesthetics of Science," *International Studies in the Philosophy of Science* 16, no. 1 (2002), 9.

34. Francis Hutcheson, *An Inquiry Concerning Beauty, Order, Harmony, Design*, 40.

35. Paul Dirac, "The Evolution of the Physicist's Picture of Nature," *Scientific American* 208, no. 5 (1963): 47.

36. Albert Einstein, *The World as I See It*, quoted by Graham Farmelo, "Foreword" to *It Must Be Beautiful: Great Equations of Modern Science*, xv.

37. James W. McAllister, *Beauty and Revolution in Science*, 160. See also 10–11, 78–81.

CHAPTER THREE

Intellectual Virtues

Intellectual virtues, also called epistemic or truth-centered virtues, include wisdom, the love of truth, intellectual honesty, intellectual courage, intellectual humility, intellectual integrity, and self-respect in intellectual matters. Intellectual virtues contribute greatly to moral creativity in science. In general, they should be counted as moral virtues. Scientific creativity is not itself a virtue but instead an achievement—the achievement of producing new and valuable products. Even so, the disposition to be creative in science qualifies as a moral virtue insofar as it serves moral ends.

An Exemplar

The life of Charles Darwin powerfully illustrates the contribution of the intellectual virtues to scientific creativity. Darwin is passionately committed to pursuing scientific understanding. He traces his success in that pursuit to the confluence of many factors but especially to his "steady and ardent" love of science.[1] The love begins during his early teenage years, develops during the five years he serves as the resident naturalist aboard the *Beagle*, and continues throughout three decades of creative work in which he publishes the *Origin of Species*, the *Descent of Man*, and a dozen additional books.

Throughout his life, Darwin focuses on the sciences that were directly relevant to the theory for which he is most famous, the evolution of species from a common ancestor by means of natural selection. No one before or after him mastered so many scientific disciplines, including geology, paleontology, comparative anatomy, developmental biology, ethnology, and psychology.[2] This mastery and focus come at a cost, including loss of his early love of poetry and drama. He reports that his "mind seems to have become a kind of machine for grinding general laws out of large collections of facts."[3] Yet there is nothing mechanical about his love of science and his passion for discovering deep explanations, all the while delighting in observation and experimentation.

Several virtues support Darwin's creativity: intellectual honesty, conscientiousness, humility, perseverance, self-respect, and courage. Honesty, to begin with, combines truthfulness and trustworthiness, both of which are rooted in respect for truth. Darwin's truthfulness is accompanied by strong independence of thought, as well as respect for evidence. In general, he is "not apt to follow blindly the lead of other men," and he endeavors to keep his "mind free, so as to give up any hypothesis, however much beloved . . . as soon as facts are shown to be opposed to it."[4] He adopts as a "golden rule" the practice of immediately making a record of criticisms of his views, to counter the natural tendency to neglect opposing evidence, and then to respond to the criticisms in subsequent publications.[5] He is also a masterful writer who patiently and with difficulty strives for clarity and care in reasoning.

As for conscientiousness, Darwin credits himself with being "superior to the common run of men in noticing things which easily escape attention, and in observing them carefully."[6] His attention to detail is illustrated by his habits of collecting. He recalls that "one day, on tearing off some old bark, I saw two rare beetles and seized one in each hand; then I saw a third and new kind, which I could not bear to lose, so that I popped the one which I held in my right hand into my mouth."[7] In this instance his conscientiousness is not altogether successful; the beetle in his mouth burns his tongue with an acrid discharge, causing him to drop two of the prized specimens.

Darwin's conscientiousness is interwoven with perseverance, shown in "unbounded patience in long reflecting over any subject" and "industry in observing and collecting facts."[8] This perseverance and industry emerge gradually. They begin with his early hobbies of collecting minerals and observing wildlife, strengthen during two years of medical school and three years in studying for the Anglican ministry, and culminate in the five-year voyage on the *Beagle*. They continue unabated throughout his adult life which, except for family activities, centered entirely on study and writing. His dedication is remarkable, as well, because his inherited wealth made it unnecessary to work for a living.

Darwin is a model of intellectual humility in being subservient to truth. He is equally humble in his self-assessments and lack of arrogance, which is noteworthy given that he is the greatest scientist of the nineteenth century and one of the most creative thinkers of all time. In a remarkable understatement, he says he possesses only "a fair share of invention and of common sense or judgment, such as every fairly successful lawyer or doctor must have, but not I believe, in any higher degree."[9] At the same time, avoiding false modesty, he is aware that his books found a large readership, suggesting his "name ought to last for a few years."[10] Although he does not seek general fame, he confesses a strong desire for recognition from other scientists. This desire supports rather than distorts his passion for discovery.[11] His intellectual humility interweaves with his ambition to be a creative scientist and to be recognized as such.

Self-respect combines reasonable self-love, self-responsibility, and self-confidence, all of which are essential to intellectual endeavors. Biographers trace Darwin's self-respect to his voyage on the *Beagle*:

> Five years and a world separated him from his old flustered, directionless, insecure self. There was a new confidence, a new earnestness; he had survived on his wits in inhospitable climes, encountered wars and savages, and trekked across the Andes. He was pleased just to be alive. He had become his own man, thinking for himself, confident enough to challenge authority.[12]

Finally, Darwin is a courageous thinker. His theory of evolution challenged widely held religious convictions, as well as dominant beliefs in biology and related sciences. His courage takes the form of intellectual daring—taking risks in pursuing a bold idea that challenges established beliefs. It also takes the form of social courage in risking derision from peers and the wider society. Admittedly his social courage has limits. Unlike Bruno and Galileo, he does not risk death from the Inquisition, but he quite reasonably fears being condemned as a heretic at a time when that condemnation mattered for his family.[13] He was ambitious but not aggressive, and he wanted to wait until the end of his life to publish what he knew would be an explosively controversial theory. It took Alfred Russel Wallace's independent discovery of evolution to provoke his completion and publication of the *Origin of Species* in 1859, and on other occasions he needed the nudge from friends to draw him into public debates.

Psychological Realism

One exemplary scientist, of course, provides no basis for generalizations about the contribution of intellectual virtues to creativity in science. Exploring that contribution should draw upon a variety of examples, including Elion, Salk, Palchinsky, and many others I discuss. Most important, it should take account of scientific studies of creative scientists. Those studies ensure psychological realism—an accurate view of human nature as a framework for moral reflection.[14] Psychological realism implies appreciating the role of the intellectual virtues in science while avoiding implausible idealizations about the purity of motives and commitments—probably few scientists are like Darwin in this regard. Psychological realism also overthrows the view that virtues are all or nothing, whether we are speaking of the scope of an individual virtue or of clusters of virtues (and vices) defining character. Above all, psychological realism avoids *moralism*, in its pejorative sense: advocating unrealistic moral rules or ideals, insensitivity to context, moral parochialism, dogmatism and rigidity, being unduly judgmental, preachy pomposity, and misusing moral judgments to gain unfair advantages.[15]

In seeking psychological realism, I draw on several leading psychologists who have developed "confluence theories" that integrate a variety of factors to explain creativity. Many of those factors are or imply intellectual virtues, although psychologists usually avoid explicitly speaking of virtues and other values. Here are some of the empirical studies relevant to Darwin's virtues.

Regarding love of truth, Robert Sternberg reports that numerous studies of creative individuals emphasize the priority of intrinsic motives, that is, motives focused on the tasks comprising the work rather than external rewards such as money and fame. Creative individuals "really love what they are doing and focus on the work rather than the potential rewards."[16] Mary Ann Collins and Teresa Amabile agree: "love for one's work is advantageous for creativity," in all domains.[17] Love means deep caring involving intrinsic motivation, that is, "the motivation to engage in an activity primarily for its own sake, because the individual perceives the activity as interesting, involving, satisfying, or challenging."[18] Creative scientists are highly motivated in their work, curious, willing to submit to the standards of excellence, and ambitious to be creative.[19]

Creative ambition is essential. To emphasize its importance, Sternberg suggests that "to a large extent, creativity is a decision."[20] The commitment to be creative arises from and in turn stimulates years of disciplined study to develop necessary intellectual skills and a strong knowledge base within a domain, as well as to develop personal thinking styles in approaching problems. The desire to be creative in science is motivated by love of understanding and by desires for achievement and recognition. Moreover, Gregory Feist observes, "the most eminent and creative scientists also tend to be more driven, ambitious, and achievement oriented than their less eminent peers."[21] Sternberg adds that they manifest a "willingness to overcome obstacles, willingness to take sensible risks, willingness to tolerate ambiguity, and self-efficacy"—elements straightforwardly linked to perseverance, courage, and self-respect.[22] ("Self-efficacy" refers to self-confidence and self-esteem as shown in conduct.)

The love of truth and the desire to make discoveries are rarely pure, in the sense of unmixed with selfish motives, and they are always mixed with elements of self-interest. An area of ongoing inquiry is the extent to which extrinsic motives such as money, fame, and power tend to support or subvert intrinsic motivation such as the love of science. Again reflecting common sense, psychologists have found that external motivation usually reinforces intrinsic motivation, although in excess it can undermine it. Competitiveness is a good example. It is an extrinsic motive in that it centers on superiority to peers, rather than on enjoyment of work itself. Especially in the contemporary world of large-scale scientific projects, competitiveness motivates most creative scientists. Another extrinsic motive, the desire for recognition, plays a major role in motivating scientific creativity.[23]

Shifting from actual motives to ideals, it would be moralistic to advocate the ideal that scientists should avoid mixed motives in pursuing truth. In fact, non-moral motives, both self-interest and caring for science or art, can strengthen moral commitment, or at least the production of moral goods. To be sure, there is a difference of degree in the motives associated with some virtues, but it is easily overstated. Consider intellectual humility, which is opposed to vanity, the vice of excessive preoccupation with what others think of us, and also opposed to arrogance, the vice of unjustifiably believing we are superior in ways that entitle us to unfair advantages. As such, intellectual humility implies keeping in check the concern for recognition, not letting it distort love for truth. Yet it would be wrongheaded and self-defeating to associate humility with indifference to honors and recognition from others.

Even Robert C. Roberts and W. Jay Wood overstate the matter when they write that intellectual humility is manifested in "an unusually low dispositional concern for the kind of status that accrues to persons who are viewed by their intellectual communities as intellectually talented, accomplished, and skilled, especially where such concern is muted or sidelined by intrinsic intellectual concerns—in particular the concern for knowledge with its various attributes of truth, justification, warrant, coherence, precision, and significance."[24] This description fits some intellectually humble individuals, but perhaps not most, at least not in every respect. Darwin is a model of intellectual humility, and he cares relatively little about how the mass of humanity views him. He cares greatly, however, about how other leading scientists regard him. Moreover, he emphasizes that his ambition for scientific recognition is closely connected with his drive to contribute to science, and that his love of science is "much aided by the ambition to be esteemed by my fellow naturalists."[25] Roberts and

Wood agree that intellectual humility can be compatible with intellectual daring and self-confidence, but they fail to appreciate how strong ambitions, often essential to creativity, typically carry with them equally strong desires for recognition. Indeed, creativity usually involves an effort to convince members of a field of study that one has make a significant discovery.[26]

Turning to courage and perseverance, there are good reasons to believe these virtues and ideals are important in contemporary science. Creative work tends to be challenging and difficult. In science, the difficulties include risks of failure, years of hard work, self-doubt, intellectual complexity, and tolerance for doubt and uncertainty. Hence, there is a need for courage and perseverance in confronting dangers and hardships. The idea of intellectual courage might sound a bit overblown as applied to most creative scientists. It is not. There are genuine risks in trying to be creative in science, especially the risk of being wrong. Large research projects are gambles that often fail to yield important results. Nobel Laureate J. Michael Bishop reports that "there is no fear in science greater than that of being wrong," and he calls for courage in confronting this danger.[27] He also emphasizes the importance of intellectual daring and bold imagination, both of which require courage. In addition, sometimes creative research is subversive in that it overthrows or forces a revision of established beliefs that shape business, government, or other social practices.

Finally, traditional moralism spawned the simplistic idea that the virtues are all or nothing, such that possessing one major virtue requires possessing others. A related idea underlies Aristotle's doctrine of the unity of the virtues, which connects with his understanding of the virtues as consistent habits or dispositions of conduct, attitude, and emotion. To the contrary, as John Dewey observed, "all character is speckled."[28] Empirical studies of motivation repeatedly show that virtues are rarely manifested in consistent ways across different domains, or even within a given area of our lives. This appreciation is reinforced by psychological studies that explore how tightly personality traits are bound to particular contexts and situations.[29] Thus, an ethics of science should emphasize that virtues manifested in creative work might not carry over into other areas of life, and even within their work scientists typically manifest moral excellence in many situations but not all.

Darwin's character was more coherent than most, and he was known for his exceptional kindness, generosity, and fair-mindedness in both his personal and professional life. In contrast, consider the Nobel Laureate physicist John Schrieffer, whom his friends describe as "a sober and cautious person." They were shocked to learn that he killed a person and seriously injured several others while driving at a 100 miles an hour, using a suspended driver's license.[30] They were even more shocked to learn that he had a long record of reckless driving, including nine speeding tickets during the past decade, and at the time of the accident his driver's license had been suspended. In addition, his impeccable honesty in doing science did not carry over to his driving: He lied to the investigating police officer about a truck that he claimed forced him off the road. Courage, honesty, and truth loving were genuine virtues in his life, but not when it came to driving. Conversely, scientists might be decent in their private lives but monstrous in their professional lives, as were many of Hitler's scientists.

Even within the context of professional activities, there might be considerable inconsistency and complexity with regard to a single virtue. Consider self-respect, which is closely related to self-esteem, a trait much studied by psychologists. Healthy self-esteem is not simply feeling good about oneself. In the literature on mental health, "self-esteem" functions as a normative term suggesting healthy, reasonable degrees of self-affirmation and self-confidence lying between the extremes of narcissism and self-denigration. In this regard, it overlaps with the moral virtue of self-respect. Self-confidence recurs in the literature on creativity as a key personality trait advancing creative endeavors.[31] More cautiously, Csikszentmihalyi finds that most creative individuals manifest a mixture of self-assurance and self-doubt, indeed a complexity of many features that are in tension with each other.[32] Self-confidence of creative people is frequently mixed with self-doubt, sometimes enormous self-doubts about their creativity ability.

Intellectual Virtues as Moral Virtues

It will be objected that intellectual virtues are not *moral* virtues at all, and hence technically not part of the ethics of science. They are virtues only in the broad sense of excellences, but not moral excellences per se. This view is traceable to Aristotle's contrast between intellectual virtues (virtues of the intellect) with moral virtues (virtues of character). For Aristotle, truth loving, intellectual honesty, intellectual humility, intellectual courage, and confidence in the pursuit of truth are excellences of the mind, not character.

In reply, Aristotle's distinction is neither clear nor compelling, and we should reject it. Aristotle distinguishes moral and intellectual virtues in terms of their functions and origins, based on an archaic division of the soul into rational and nonrational parts.[33] The rational part is governed by intellectual virtues learned during education. The nonrational part, which includes emotions and bodily desires, is governed by moral virtues acquired as habits during early development. But this dichotomy is a muddle. For one thing, the excellences of the intellect involve the emotions; for example, delight in truth, contempt for shoddy thinking, and hope for new discoveries. In addition, all excellences require habit formation and teaching from childhood and throughout adulthood. For another thing, as even Aristotle emphasized, intellectual virtues are thoroughly interwoven with other moral virtues, not only in providing guidance but in partly defining them. For example, we cannot be temperate, courageous, or fair without exercising practical wisdom (*phronesis*), that is, knowledge of what is morally good, which Aristotle viewed as the primary intellectual virtue. In any case, virtues like intellectual honesty and intellectual humility are clearly moral virtues in any contemporary sense of "moral." They are desirable traits of character that tend to contribute to the good of both ourselves and others.

Thinking in this vein, Linda Zagzebski insightfully argues that "intellectual virtues ought to be treated as a subset of the moral virtues."[34] Morality is not solely about respecting others' rights and promoting their good. It concerns as well our dealings with truth and with "the normative aspects of cognitive activities."[35] To be sure, we can continue to speak of intellectual virtues, so long as we understand them as those moral virtues directly concerned with truth, belief, and knowledge. For example, honesty is a moral virtue that concerns truthfulness in all areas of life, but we can distinguish honesty in the search for significant truth (intellectual honesty), honesty within a marriage (marital honesty), and honesty with one's therapist (therapeutic honesty). Again, courage is the moral virtue of admirable risk taking in all areas of life, but we can distinguish between the courageous pursuit of truth (intellectual courage), courage in confronting risks to life and limb (physical courage), courage in confronting social criticism (social courage), and courage in undertaking investment risks (financial courage). The relevant contrast is not between the intellectual and the moral, but instead among various dimensions of life in which moral virtues play a role.

What about scientific creativity—is it a moral virtue too? Zagzebski believes it is. She suggests that creativity, in science and elsewhere, is a supererogatory moral virtue in that it goes beyond what can be expected of most people.[36] Here we need to be careful. As I defined it, creativity is an achievement rather than a virtue—the achievement of making new and valuable discoveries or inventions. In general, we can distinguish two kinds of excellence: achievements and virtues. As an analogy, consider two senses of athletic excellence: athletic achievement and athletic prowess. Winning Wimbledon is an outstanding achievement in tennis, and the achievement manifests virtues such as self-control and hard work, as well as talent and some luck. Likewise, the excellence of making a creative scientific discovery is distinct from the excellent traits in the scientist that help make it possible. Excellence in achievement and excellence as traits are closely connected, but distinct.

Perhaps, however, Zagzebski had in mind personality traits such as the tendency to make creative achievements, together with the disciplined striving to be creative. Such a tendency is a trait rather than an isolated achievement. Thus, when individuals like Charles Darwin and Gertrude Elion embrace creativity as a personal ideal, pursue it deliberately and habitually, and then make significant discoveries, they manifest an intellectual virtue. In turn, that intellectual virtue is a moral virtue when it is connected with moral goods, as it is with Elion (saving lives) and Darwin (understanding our place in the universe).

An alternative view is to think of scientific creativity and certain other intellectual excellences as "dependent" moral virtues—excellences that count as moral when they make (or tend to make) moral contributions.[37] Here two points are important. First, scientific creativity and other intellectual virtues do tend to make moral contributions, directly or indirectly—that is a central theme developed throughout this book. Second, the same qualms about counting intellectual virtues as moral virtues can arise with many other moral virtues. Is the suicide bomber who willingly dies for his cause courageous, and is the Nazi guard who diligently performs his duties conscientious? Or instead, do the ideas of courage and conscientiousness apply only in cases where morally desirable ends are involved, making them dependent virtues? If we make the latter choice, virtually all moral virtues would be dependent virtues. That way of talking is not implausible. I find it more plausible and straightforward, however, to accept the intellectual virtues as moral virtues that, like all moral virtues, are interwoven with other virtues that place limits on them.

MacIntyre's Framework

We can now place the intellectual virtues in a wider perspective on virtue-oriented ethics set forth by Alasdair MacIntyre in *After Virtue*. MacIntyre characterizes the sciences, like other professions, as social practices: cooperative human activities centered on distinctive internal goods that define the practice. For example, health is the internal good in medicine, justice in law, empirical understanding in (theoretical) science, and safe and useful products in engineering. Internal goods also include valuable ways of life, together with the enjoyments they make possible for participants in social practices. In science, internal goods include meaningful ways of life for scientists who develop, promulgate, and apply scientific understanding—as well as contributions to meaningful lives for members of the public who participate in science indirectly. Both types of internal goods are public goods, in that they contribute to the good of the community.

Virtues are excellences of character that enable participants to pursue the internal goods of a social practice with excellence. For example, honesty, courage, self-discipline, integrity, fairness, and team play contribute to meeting the standards of excellence in all practices, and thereby promote the practices' internal goods. The virtues also enable individuals to live good lives, which we can think of as both morally desirable and personally satisfying.[38] MacIntyre avoids any sharp dichotomy between moral and intellectual virtues. For example, he specifies that honesty is a moral virtue that implies not only trustworthiness but also "carefulness for the facts," and courage implies a willingness to "take whatever self-endangering risks are demanded" in pursuing the internal goods of the practice.[39] The honesty and courage of Darwin helped make possible both the discovery of significant truths and sustained his meaningful life in the pursuit of those truths.

In the sciences, as in the arts, creativity is the highest form of excellence. Science advances by discovering new and valuable understanding through creative "attempts to sustain progress and to respond creatively to problems."[40] Creativity comes in many degrees, but exceptional creativity transforms an entire social practice, as Darwin did in revolutionizing biology and related fields. In addition, as standards of excellence are met, so as to advance the internal goods of a practice, "human powers to achieve excellence, and human conceptions of the ends and goods involved, are systematically extended."[41]

MacIntyre focuses on cardinal virtues such as wisdom, temperance, courage, honesty, and justice, which promote excellence in all social practices. In addition, each social practice makes distinctive demands that give priority to some virtues. As Jacob Bronowski emphasizes, the values of science are not mere instantiations of a universal code of conduct, but instead "have grown out of the practice of science, because they are the inescapable conditions for its practice."[42] He regards the "habit of truth" as the paramount value in science, by which he means the commitment to a process of seeking truth, along with values such as independent and original thought that sustain that process.

Finally, I should note that MacIntyre conflates two different criteria in distinguishing internal and external goods.[43] According to one criterion, internal goods are public (community) goods, and external goods are privately owned (individual or institutional) goods. Thus, scientific understanding is a public good, whereas the salary earned by a scientist is an external good. According to a second criterion, internal goods are those goods defined by specific practices, in contrast with "external goods" that can be acquired in virtually any practice. Thus, the internal good of scientific understanding is defined using scientific concepts, whereas the external goods of money, power, and fame are not defined by reference to science or any other specific practice. MacIntyre fails to appreciate that the application of this second criterion depends entirely on the level of abstraction in specifying goods. "Recognition as a creative scientist" and "fame as a scientist" are defined in terms of a particular practice, and hence are internal goods. Yet, they are also instances of recognition and fame, which can be found in all practices, and are hence external goods when described in generic terms. In short, they can be either internal or external goods, depending on how abstractly or concretely they are specified. It is important to bear this point in mind when thinking about the goods defining science and the motivation of scientists.

To conclude, professional ethics in science includes good character and the ideal of being creative, in addition to the responsibilities set forth in various codes of ethics for scientists. At the same time, the relation between character and creativity is complex. Good character can contribute greatly to science, but it is obviously not sufficient for being creative. Talent and luck might play more prominent roles than good character. Equally important, non-moral motives contribute much to creativity, and thereby to morality, as discussed next.

Notes

1. Charles Darwin, *The Autobiography of Charles Darwin*, ed. Nora Barlow (New York: W.W. Norton, 1958), 141. Although I quote from his autobiography, Darwin's self-assessment is confirmed by numerous biographers.
2. Niles Eldredge, *Darwin: Discovering the Tree of Life* (New York: W.W. Norton, 2005), 91–92.
3. Darwin, *Autobiography*, 139; cf. 109.
4. Darwin, *Autobiography*, 141.
5. Darwin, *Autobiography*, 120, 123.
6. Darwin, *Autobiography*, 141.
7. Darwin, *Autobiography*, 62.
8. Darwin, *Autobiography*, 145.
9. Darwin, *Autobiography*, 140.
10. Darwin, *Autobiography*, 139.
11. Darwin, *Autobiography*, 141.
12. Adrian Desmond and James Moore, *Darwin* (New York: W.W. Norton and Company, 1991), 196.

13. Desmond and Moore, *Darwin*, 296.

14. Owen Flanagan, *Varieties of Moral Personality: Ethics and Psychological Realism* (Cambridge, MA: Harvard University Press, 1991), 32.

15. Cf. Haavard Koppang and Mike W. Martin, "On Moralizing in Business Ethics," *Business and Professional Ethics Journal* 23, no. 3 (2004): 107–14.

16. Robert J. Sternberg, *Wisdom, Intelligence, and Creativity Synthesized* (New York: Cambridge University Press, 2003), 108.

17. Mary Ann Collins and Teresa M. Amabile, "Motivation and Creativity," in *Handbook of Creativity*, ed. Robert J. Sternberg, (New York: Cambridge University Press, 1999), 308.

18. Collins and Amabile, "Motivation and Creativity," 299.

19. Gregory J. Feist, "The Influence of Personality on Artistic and Scientific Creativity," in *Handbook of Creativity*, ed. Robert J. Sternberg, 280.

20. Robert J. Sternberg, *Wisdom, Intelligence, and Creativity Synthesized*, 106.

21. Gregory J. Feist, "The Influence of Personality on Artistic Creativity," 280.

22. Robert J. Sternberg, *Wisdom, Intelligence, and Creativity Synthesized*, 108.

23. Collins and Amabile, "Motivation and Creativity," 305.

24. Robert C. Roberts and W. Jay Wood, "Humility and Epistemic Goods," in *Intellectual Virtue*, ed. Michael DePaul and Linda Zagzebski, (Oxford: Clarendon Press, 2003), 271.

25. Darwin, *Autobiography*, 141; cf. 81–82.

26. Mihaly Csikszentmihalyi, *Creativity* (New York: Harper Collins, 1996), 42.

27. J. Michael Bishop, *How to Win the Nobel Prize* (Cambridge, MA: Harvard University Press, 2003), 62.

28. John Dewey, *Human Nature and Conduct* (New York: Modern Library, 1957), 47.

29. John M. Doris, *Lack of Character: Personality and Moral Behavior* (New York: Cambridge University Press, 2002).

30. The "accident" occurred in September, 2004, and in August, 2005, he pleaded guilty to manslaughter. Steve Chawkins, "The Baffling Descent of a Nobel Prize Winner, *Los Angeles Times* (August 13, 2005), B1, B7.

31. Feist, "The Influence of Personality on Artistic Creativity," 280.

32. Csikszentmihalyi, *Creativity*, 69.

33. Aristotle, *Nicomachean Ethics*, trans. W. D. Ross, in *Basic Works of Aristotle* ed. Richard McKeon (New York: Random House, 1941), 1103a3–7 and 1139a.

34. Linda Trinkaus Zagzebski, *Virtues of the Mind* (New York: Cambridge University Press, 1996), 139.

35. Zagzebski, *Virtues of the Mind*, 255. Zagzebski says intellectual virtues are self-regarding (255–66), but we do better to call them truth-regarding virtues that contribute to the well-being of both oneself and others.

36. Zagzebski, *Virtues of the Mind*, 125, 155.

37. Michael Slote discusses dependent virtues in *Goods and Virtues* (Oxford: Clarendon Press, 1983), 62.

38. Cf. John Kekes, *The Art of Life* (Ithaca: Cornell University Press, 2002).

39. Alasdair MacIntyre, *After Virtue*, 2d ed. (Notre Dame, IN: University of Notre Dame Press, 1984), 191.

40. MacIntyre, *After Virtue*, 190.

41. MacIntyre, *After Virtue*, 187.

42. Jacob Bronowski, *Science and Human Values* (New York: Harper and Row, 1965), 60.

43. Cf. Mike W. Martin, *Meaningful Work: Rethinking Professional Ethics* (New York: Oxford University Press, 2000), 125–27, 184–86.

CHAPTER FOUR

Paradoxes of Motivation

In suggesting that "philanthropy is almost the only virtue which is sufficiently appreciated by mankind," Thoreau did not wish to denigrate charity, but he took offense when even minor Christian leaders were ranked above Newton, Shakespeare, and other creative giants "who by their lives and works are a blessing to mankind."[1] Such individuals might be motivated primarily by caring for goods that are usually considered non-moral, such as scientific truth, aesthetic appreciation, and creative achievement. Yet, paradoxically, these creative individuals often benefit humanity far more than they could have through direct moral or philanthropic service. This *creativity paradox* has been given less attention than paradoxes of self-interest and altruism, but it contains an important insight about moral motivation. It also forces us to clarify the distinction between moral and non-moral goods, and hence our conception of what morality is.

Self First, Others First, Non-Moral First

A paradox is a seeming contradiction, incongruity, or irony that conveys an important truth. The paradoxes of moral motivation are not logical conundrums but instead empirical complexities surrounding moral motivation. Typically they are invoked in hortatory settings, but they have implications for moral psychology and ethical theory.

Most paradoxes of moral motivation turn on the interplay of self-interest and altruism.[2] Thus, *self-first paradoxes* identify how self-seeking contributes to the well-being of others. For example, to love others we should first love ourselves, and to make others happy we do well to begin by being happy, for only in these ways do we have the appropriate gifts to offer. The most famous self-first paradox concerns Adam Smith's metaphor of an invisible hand, whereby self-seeking in the marketplace benefits others more than direct and systematic altruism could. In particular, the self-interested motives of entrepreneurs (money, influence, recognition) lead them to create jobs for workers, quality products for consumers, profit for investors, and wealth for philanthropists.[3] Systematic altruism in the marketplace would be less effective in creating these benefits of competitive capitalism, both because self-interest is a stronger motive than altruism and because we understand our interests better than those of others.

Reversing direction, *others-first paradoxes* identify how contributing to the good of others is sometimes the best way to advance our self-interest. For example, self-sacrifice promotes self-development, and we find ourselves by losing ourselves in service to others. The most familiar others-first paradox is the happiness paradox: to get happiness we should forget about it and instead make others happy. Another example is the freedom paradox: submission liberates, and freedom is won by surrendering it to worthy causes. The idea is that loyalty to spouses, community, or God opens new resources for growth, including the goodwill of others. This notion is also present in twelve-step recovery programs: "We admitted we were powerless over alcohol—that our lives had become unmanageable," and we "came to believe that a Power greater than ourselves could restore us to sanity."[4]

In contrast, *non-moral-first paradoxes* turn on the interplay of moral and non-moral goods rather than self-interest and altruism. In such paradoxes we identify ways in which pursuing non-moral goods contributes to moral goods, for others and for ourselves. An example is John Stuart Mill's expanded statement of the happiness paradox: "those only are happy . . . who have their minds fixed on some object other than their own happiness; on the happiness of others, on the improvement of mankind, even on some art or pursuit, followed not as a means, but as itself an ideal end."[5] The pursuit might be in the sciences, arts, athletics, work, or any number of other endeavors involving non-moral goods (in addition to moral ones). It is important that these goods be cared about for their own sake, as "ideal ends," in order for them to provide the deep and lasting enjoyments which contribute to our happiness.

Henry Sidgwick makes an even stronger claim by inverting the happiness paradox. He says that "the fullest development of happy life for each individual seems to require that he should have other external objects of interest besides the happiness of other conscious beings," and that we should engage in the "disinterested pursuit" of "Virtue, Truth, Freedom, Beauty, etc., *for their own sakes*."[6] The moral goal of maximizing overall happiness requires pursuing non-moral goods as desirable in their own right, rather than as mere means to moral ends. The freedom paradox can be recast in a similar manner by saying that we are morally liberated through submission to non-moral goods such as truth, beauty, and creative striving. In addition, Adam Smith's invisible-hand paradox can be recast along these lines by noting that entrepreneurs contribute to the moral well-being of themselves and others when they value the non-moral excellence of their products and services.

Non-moral-first paradoxes, then, overlap with the self-first and others-first paradoxes. In particular, the happiness paradox suggests we are liberated from anxious self-absorption by focusing outward, whether by caring for other people or for non-moral goods.[7] This overlap among the paradoxes derives from the intertwining of goods for ourselves, goods for others, and moral and non-moral goods, as well as the typical intertwining of our motives aimed at such goods. Accordingly, the paradoxes should not be construed as polarizing these categories of goods. Instead, the paradoxes identify ironic ways in which pursuing one category of goods requires focusing on another category of goods, for their sake.

The Creativity Paradox

The creativity paradox is the most dramatic non-moral-first paradox. According to it, creative endeavors in the pursuit of non-moral goods benefit others morally, and they also contribute to the meaning and happiness creative individuals find in life. The idea is expressed by one of H. G. Wells's characters: "the motive that will conquer cancer will not be pity nor horror; it will be curiosity to know how and why. . . . Desire for service never made a discovery."[8] In general, although compassion and altruism enter into the justification of science, they are not the primary motives of most scientists. Hence the creativity paradox: non-moral motives of creative individuals often lead to greater moral contributions than do altruistic motives.

J. Michael Bishop, a Nobel Laureate for his cancer research, sympathetically cites the remarks of Wells's character. By no means does Bishop endorse an amoral view of science. On the contrary, he underscores the importance of honesty, courage, fairness, collaboration, and mentoring of junior colleagues, all of which serve to advance scientific inquiry.[9] In the main, however, he portrays moral values as constraints on how inquiries are conducted, rather than primary motives for inquiry. The primary motivation comes from the non-moral desires of curiosity, truth-seeking, striving to be creative, and appreciation of nature's beauty.

Although not moral per se, at least as usually understood, these motives and their accompanying ideals and character traits are widely admired by scientists and laypersons alike. Thus, we are moved when we read of the heroic sacrifices of Marie Curie in studying radioactivity and discovering polonium and radium. Not only does she confront special obstacles as a woman in a male scientific world, but she risks her health and leads a stoical life in the service of science. To be sure, she also makes direct humanitarian contributions, including designing and overseeing the operation of twenty mobile X-ray units and some two hundred more stationary units to treat wounded soldiers during World War I. Nevertheless, curiosity and joy in making discoveries are her primary motives in doing scientific research.[10] She and her husband Pierre have little interest in general fame, although they desire acknowledgment from their peers for their scientific contributions. Nor do they care about great wealth, and they choose not to patent the industrial processes they developed.

Marie Curie is disinterested (unbiased, objective) in her pursuit of truth but also passionate. Her fascination with math and science become inseparable from her self-identity. Moreover, she does not merely want to make discoveries; she wants to be the first person to make them—a desire typical of creative scientists. Considerably less typical, however, is her indifference to financial success and fame. Hence we should immediately add an example of where desires for worldly goods are mixed with scientific commitments.

The discovery of the structure of DNA by James Watson and Francis Crick contributes enormously to human well-being. It fundamentally advances understanding, makes possible new drugs that save countless lives, and paves the way for eventual gene therapies. The motives of Watson and Crick are not altruistic, however, at least according to Watson's account in *The Double Helix*. On the one hand, their motives center on scientific goods, which include the pursuit of understanding and also aesthetic appreciation of nature. Watson expresses his "delight and amazement" in the "profoundly interesting" discovery of a structure that is "too pretty not to be true."[11] On the other hand, they are obsessed with being the first to solve the riddle of DNA, in order to win a Nobel Prize and have their names forever associated with the double helix structure. An additional motive comes from their rivalry with Linus Pauling in the race to discover DNA's structure.

Watson and Crick proceed in a spirit of scientific adventure, personal ambition, and competitive rivalry, not moral commitment. Morality is mentioned only twice in Watson's memoir. In the preface he states that the scientific world is "complicated by the contradictory pulls of ambition and the sense of fair play," and in the epilogue he expresses regret that he and Crick failed to appreciate the virtues of Rosalind Franklin, whose X-ray crystallography had been crucial to their discovery—and used without her permission.[12] Nevertheless, their discovery is morally momentous. In this way, they exemplify the non-moral-first paradox: non-moral motivation can lead to far greater moral contributions than direct moral motivation.

Their endeavors also exemplify self-first paradoxes, in that their self-seeking contributes to the community. In general, creativity and other non-moral paradoxes overlap with self-first paradoxes, reflecting how self-interest and scientific curiosity typically interweave in the lives of committed scientists. Robert K. Merton draws attention to this mixing of motives in underscoring that competition is connected with scientific originality in making a discovery first: "Recognition and fame thus appear to be more than merely personal ambitions. They are institutionalized symbol and reward for having done one's job as a scientist superlatively well."[13]

Watson did not seek fame in the abstract. He sought to achieve fame *as a scientist*. The relevant fame is not a good external to science, as if it could be achieved by participating in just any social practice. For this reason, as noted in chapter 3, I rejected Alasdair MacIntyre's way of distinguishing internal and external goods of social practices.[14] MacIntyre says internal goods are public goods that are defined in terms of a given practice and recognizable only in terms of the practice. In science, they include scientific understanding and the meaningful lives made possible through being a scientist. External goods are goods which constitute private property and can be achieved through any practice, for example, fame, power, and money. MacIntyre overlooks, however, that goods can be described at various levels of generality.[15] When an individual's motive is to achieve fame as a scientist, fame becomes an internal good that can only be achieved through science.

The discussion so far suggests there is no great mystery behind the creativity paradox. Science, like many other social practices, contributes to moral goods, but individuals can participate in science with myriad motives. Typically, self-interest intertwines with practice-specific motives such as a desire to pursue scientific truth and to be creative. Direct moral motivation, such as to help others through our work, might not be present in a given scientist, and yet that scientist might contribute greatly to the well-being of others. In tune with the creativity paradox, the drive for creative excellence in science often brings about greater moral benefits than does direct humanitarian service.

Dangerous Paradox

The paradoxes of moral motivation are themselves used to motivate, guide, and justify. Thus, from Adam Smith to Ayn Rand, libertarians have used the invisible-hand argument to defend laissez-faire capitalism that celebrates economic self-seeking with minimum government regulation. Again, Mill's happiness paradox enters into his defense of a particular version of utilitarianism, which emphasizes "higher pleasures" attached to valuable activities such as scientific inquiry. As well, the freedom paradox is a staple in homiletic and hortatory contexts, such as religions advocating particular value perspectives.

There is a danger in all this, since invariably the paradoxes are one-sided and eclipse other important truths. The one-sidedness should be obvious in light of the directly opposing directions taken by the self-first paradoxes and others-first paradoxes. In particular, the self-first paradox can be used to validate constricted, callous, and selfish lives. Again, Adam Smith's invisible-hand paradox fails to take account of the harmful side effects of greed, especially harm to the environment in the form of pollution, depletion of scarce resources, and endangerment of species, all of which are harms that a free market generates and which require government intervention of the sort opposed by libertarians. Conversely, the others-first paradox can be used to endorse endeavors based on misguided altruism, imprudence, and excessive generosity in giving away resources needed for ourselves and our family. Clearly, interpreting and applying the paradoxes requires good judgment.

The creativity paradox is equally one-sided, in ways that render it treacherous as a guide and a justification. In the strong version set forth by Wells's character, the creativity paradox is hyperbole. It is patently false that a desire for service never led to a discovery, and that all creative cancer researchers are motivated exclusively by scientific curiosity. Recall Gertrude Elion, who entered her scientific career in direct response to the slow and painful death of her grandfather, and who expresses joy in developing drugs that help children with leukemia. Plainly, it was her training as a chemist that enabled her to do scientific research, and no doubt that training could not have succeeded unless she also derived pleasure from the pursuit of truth. But moral motives of service and compassion drove her research. In general, the creativity paradox eclipses the importance of moral motives in the lives of scientists and other creative individuals.

For this same reason, Watson's memoir cannot be taken as a model for the motives of all scientists. Indeed, some scientists were shocked by the prominence Watson gives to self-seeking motives. The distinguished biologist

Robert Sinsheimer, later renown for being the catalyst for the human genome project, calls the book "saddening" and Watson's world "unbelievably mean in spirit," a world that eclipses "the humane and esthetic qualities of science" by focusing exclusively on "intense ambition—for the mundane prize, not the advancement of truth nor the service of humanity."[16] Sinsheimer expresses concern about the potentially negative effect of the book on high school students who upon reading it come to see the discovery of the structure of DNA as a crass exercise in egoism. I think Sinsheimer overstates his point. He fails to appreciate how Watson conveys the excitement of scientific inquiry, as well as his collegial collaboration with Crick and his willingness to cooperate with peers within a wider context of competition and rivalry in making discoveries.[17] Even so, Sinsheimer reminds us that the motives of scientists generally include moral commitments. Indeed, Watson himself cautions against stereotyping the motives of scientists, pointing out that "styles of scientific research vary almost as much as human personalities."[18] The creativity paradox eclipses the importance of moral motives and ideals in science and other social practices.

This eclipse encourages narrow compartmentalization of scientific inquiry, of the sort that insulates it from moral scrutiny by its participants. There is already a strong temptation for scientists and engineers to suspend moral scrutiny in pursuing their work, whether owing to self-interest or to the sheer excitement of immersing themselves in technical work. Indeed, there is a large area where it is perfectly acceptable to focus on narrow problems in order to make progress in solving puzzles. Moreover, allowing moral values to influence inquiries might distort the search for truth, especially when the values are not justified, as occurred in much Soviet science. Nevertheless, to compartmentalize risks transforming the pursuit of truth into a Faustian bargain that promotes evil rather than good, as I discuss more fully in chapter 6.[19]

Richard P. Feynman cautions against compartmentalization in his reflections on the Manhattan Project. His original motivation and justification for participating in building the first atomic bomb was to defeat Hitler, as well as to prevent the Nazis from making an atom bomb first. For several years he was immersed in the technical aspects of the work. He expresses regret, however, that when Nazi Germany surrendered he was so absorbed in his work that he did not pause to reassess his participation. The experience taught him that "if you have some reason for doing something that's very strong and you start working at it, you must look around every once in a while and find out if the original motives are still right."[20] Conscience, as well as curiosity and creativity, should shape the work of scientists.

Moral and Non-Moral

Our understanding of the creativity paradox, and other non-moral-first paradoxes, turns on how we distinguish between moral and non-moral goods, and hence on our conception of morality. So does our ability to integrate moral and non-moral values in order to minimize the dangers of compartmentalizing. We have been considering an informal, everyday conception of morality that includes direct concern for the well-being of ourselves and others; in contrast, non-moral goods include such things as truth, beauty, and creative achievement. We might, however, narrow or expand this everyday conception. One popular option is to narrow morality by limiting it to relationships with other people, thereby establishing a dichotomy between morality and prudence. Such narrowing occurs in Harry Frankfurt's otherwise insightful discussions of the freedom paradox. Frankfurt avers that "morality is most particularly concerned with how our attitudes and our actions should take into account the needs, the desires, and the entitlements of other people."[21] Using this narrow conception, which makes our own good a non-moral matter of prudence, Frankfurt suggests that morality is vastly overrated as a guide to our conduct. Meaningful lives are far more dramatically shaped by disinterested but passionate caring about various non-moral goods, such as goods in science and the arts, as well as about specific individuals, rather than by general duties to others. Frankfurt's conception of morality amplifies the non-moral-first paradox by simultaneously increasing the role of non-moral goods and narrowing the domain of morality to universal duties to others.

In a somewhat related vein, Bernard Williams conceives morality narrowly as a special system of values centered on obligations, free will, and blame and guilt. He contrasts morality with "ethics," defined as the entire domain of values that give life meaning.[22] In opposition to impartial morality, narrowly conceived, Williams sets forth an ethics of character centered on the virtues, relationships with particular individuals and specific groups, and personal (morally optional) projects in promoting non-moral goods. He understands character as a nexus of "ground projects" that might include moral commitments involving compassion and justice or relationships of friendship and love, but that might also be non-moral activities such as work in science, art, or athletics.[23]

This idea of personal activities is especially illuminating in thinking about creative endeavors in science. Such endeavors have ethical significance insofar as they enter into fulfilling lives of scientists and the community, but they can be defined without invoking moral values. Even so, most personal activities and relationships interweave with moral commitments, in various degrees, as well as with non-moral and self-interested commitments. A life in science, for example, might be motivated by blended desires for personal meaning and benefits to others, both of which are moral aims (in my view). It might even be driven primarily by direct humanitarian motives, as with Gertrude Elion.

The idea of personal projects that interweave moral and non-moral values permeates modern German thought. It connects with themes of authenticity, self-realization, and the desirability of pursuing our own way of bringing value into the world.[24] It is an idea, for example, central to Goethe's emphasis on "projects of love" that enrich our lives and others. As Albert Schweitzer comments in summarizing Goethe's outlook: "Everyone must realize the love that is peculiar to him," and "We have to realize the good that is part of our personal being, thus perfecting our personality, not everyone in the same way, but each as an ethical being in his own right."[25] This love includes the love of science, which is frequently blended with humanitarian as well as self-interested motives.

Turning to the second option, we might expand morality to include many non-moral goods. For example, some versions of utilitarianism widen the list of intrinsic goods that right action promotes. Instead of restricting intrinsic goods to happiness or pleasure, so-called "ideal utilitarianism" embraces a plethora of additional goods such as knowledge, appreciation of beauty, love, friendship, and the virtues. If we add creative accomplishment to this list, the connection between morality and non-moral goods is tightened even further. In this view, Curie, Watson and Crick, and Elion are all engaged in a moral enterprise of seeking truth and creative expression. For utilitarians their conduct will need to be assessed in terms of overall production of goods. Given the dramatic impact of their achievements, however, it would seem plausible to portray their scientific activities as forms of right conduct. In this connection, James Rachels, inspired by Sidgwick, recommends a "multiple-strategies utilitarianism" in which we promote the general good through direct concern for many more specific goods, moral and non-moral.[26]

Another expansion of morality, this time within virtue ethics, abandons Aristotle's distinction between moral and intellectual virtues, where intellectual virtues include curiosity, love of truth, drive to be creative, unbiased judgment, clarity of thought, attention to detail, disciplined response to evidence, openness to new hypotheses, and appreciation of elegant and beautiful equations. Instead of regarding these virtues of the mind as non-moral virtues, we should construe them as moral virtues. They are intellectual virtues only in that they are moral virtues pertaining to intellectual endeavors. That is the view I endorsed in chapter 3, following the lead of

Linda Zagzebski who, echoing Spinoza and Hume, insists that intellectual virtues are special instances of moral virtue.[27]

Zagzebski's broad conception of morality might seem to dissolve the creativity paradox by collapsing non-moral matters into moral matters. In fact, her conception of morality merely prompts a revision of the paradox. Thus, a contrast remains between the moral import of the self-fulfillment of scientists and the contribution scientists make to others. The creativity paradox is recast in terms of how exercising the moral (intellectual) virtues in the pursuit of scientific understanding indirectly contributes to other moral goods such as the meaningful lives of scientists and the good of the wider community.

In sum, the non-moral-first paradox, and especially the creativity paradox, highlights how concern for non-moral goods, for their sake, often contributes dramatically to moral goods. It reminds us that moral and non-moral goods are more tightly interwoven in human motivation than ethical theories usually allow. Ethical theorists do well to take fuller account of how caring for non-moral goods, for their sake, contributes to morality. In addition, we need to take account of the role of luck, which is our next topic.

Notes

1. Henry David Thoreau, *Walden and Civil Disobedience* (New York: Penguin, 1983), 120.
2. Mike W. Martin, *Virtuous Giving* (Bloomington: Indiana University Press, 1994), 151.
3. Adam Smith, *An Inquiry into the Nature and Causes of the Wealth of Nations* (New York: Oxford University Press, 1976), 456.
4. *Alcoholics Anonymous*, 3d edition (New York: Alcoholics Anonymous World Services, 1976), 59.
5. John Stuart Mill, *Autobiography* (New York: Penguin Books, 1989), 117.
6. Henry Sidgwick, *The Methods of Ethics*, 7th edition (Chicago: University of Chicago Press, 1962), 405–406. See also Robert Adams, "Motive Utilitarianism," *Journal of Philosophy* 73 (1976): 467–81.
7. Cf. Bertrand Russell, *The Conquest of Happiness* (New York: Liveright, 1930), 188.
8. H. G. Wells, *Meanwhile* (London: Ernest Benn Limited, 1927), 37. See also G. H. Hardy, "A Mathematician's Apology," in *Genius and Eminence*, 2nd ed., ed. Robert S. Albert (New York: Pergamon Press, 1992), 394.
9. See J. Michael Bishop, *How to Win the Nobel Prize* (Cambridge, MA: Harvard University Press, 2003), 60–61.
10. See Eve Curie, *Madame Curie*, trans. Vincent Sheean (New York: Doubleday, Doran & Company, 1938), and Susan Quinn, *Marie Curie: A Life* (Reading, MA: Perseus, 1995).

11. James D. Watson, *The Double Helix*, in *The Double Helix*, ed. Gunther S. Stent (New York: W.W. Norton, 1980), 108, 124. See also 116.

12. Watson, *The Double Helix*, 3, 132.

13. Robert K. Merton, "Making It Scientifically," in *The Double Helix*, ed. Gunther S. Stent, 216. See also David L. Hull, *Science as a Process* (Chicago: University of Chicago Press, 1988), 304; and Peter Godfrey-Smith, *Theory and Reality* (Chicago: University of Chicago Press, 2003), 99.

14. Alasdair MacIntyre, *After Virtue*, 2nd ed. (Notre Dame, IN: University of Notre Dame Press), 188–90.

15. Cf. Mike W. Martin, *Meaningful Work: Rethinking Professional Ethics* (New York: Oxford University Press, 2000), 185.

16. See Robert L. Sinsheimer, "*The Double Helix*," in *The Double Helix*, ed. Gunther S. Stent, 191–94.

17. Cf. Kay Redfield Jamison, *Exuberance: The Passion for Life* (New York: Alfred A. Knopf, 2004), 190–95.

18. Watson, *The Double Helix*, 3.

19. Arnold Pacey, *Meaning in Technology* (Cambridge, MA: MIT Press, 1999), 171–98.

20. Richard P. Feynman, *The Pleasure of Finding Things Out* (Cambridge, MA: Perseus, 1999), 231–32.

21. Harry G. Frankfurt, *The Reasons of Love* (Princeton, N.J.: Princeton University Press, 2004), 7.

22. Bernard Williams, *Ethics and the Limits of Philosophy* (Cambridge, MA: Harvard University Press, 1985).

23. Bernard Williams, "Persons, Character and Morality," in Bernard Williams, *Moral Luck* (New York: Cambridge University Press, 1981), 13.

24. Charles Taylor, *The Ethics of Authenticity* (Cambridge, MA: Harvard University Press, 1992).

25. Albert Schweitzer, *Goethe: Five Studies*, trans. Charles R. Joy (Boston: Beacon Press, 1961), 139–40.

26. James Rachels, *The Elements of Moral Philosophy*, 4th ed. (New York: McGraw-Hill, 2003), 198.

27. Linda Trinkaus Zagzebski, *Virtues of the Mind* (New York: Cambridge University Press, 1996), xiv. See also 137–65.

C H A P T E R   F I V E

Serendipity

The term *serendipity* was coined in the eighteenth century by Horace Walpole, who was inspired by a fairy tale called the "Three Princes of Serendip": "as their Highnesses traveled, they were always making discoveries, by accidents and sagacity, of things which they were not in quest of."[1] The accidents were good luck, where luck refers to benefits (or harms) from events we cannot be expected to foresee and control.[2] The sagacity includes observations and inferences that lead to significant discoveries. This confluence of good luck and insightful response frequently occurs in science. The confluence also bears on issues about responsibility connected with moral luck: that is, luck relevant to moral evaluations of conduct and character. I discuss credit for scientific contributions, then turn to blame for wrongdoing, and conclude with luck in making career decisions.

Credit for Contributions

Alexander Fleming's discovery of penicillin is the most famous example of scientific serendipity.[3] According to his account, in 1928 he is doing routine laboratory work on influenza. Upon returning from a summer vacation, he notices an unusual clear area in a Petri dish containing a culture of bacteria used in an experiment completed weeks earlier. Instead of discarding or washing the dish, he investigates. Apparently, a speck of mold entered the dish when it was briefly uncovered or improperly sealed, and the vacation provided time for the mold to grow. After identifying the mold as from the genus *Penicillium*, he names the antibiotic substance penicillin. At the time he could not have guessed that his discovery would save millions of lives. Although he devotes several years to studying penicillin, a decade passes before other researchers, pressed by the need to help injured soldiers during World War II, manage to develop penicillin for use as an antibiotic.

Historians suggest that Fleming's account is influenced by poor memory and fallacious inference.[4] Fleming writes his account of the discovery in 1944, after he becomes famous, which allowed sixteen years for memories to dim. Indeed, Fleming's lab books first mention the discovery nearly two months after his return from vacation, and they record an experiment done at that time, not earlier. Most likely, Fleming was doing a different set of experiments than he recalls. Moreover, the stray spore came not from an open window, for the laboratory windows were almost always kept sealed, but instead up the stairwell from a lab on the floor below him where another scientist was working with molds.

Either way, Fleming's discovery is made possible by a remarkable chance encounter between one among thousands of different fungi and one among thousands of different bacteria. The laboratory events surrounding this encounter are an instance of *circumstantial luck*: luck in the context of action, including luck in the options we have and the decisions we must make.[5] In addition, Fleming has *outcome luck*: luck pertaining to consequences or results, in that penicillin has the unforeseen efficacy as a broadly efficacious antibiotic with few bad side effects. Because Fleming was not trying to discover a general antibiotic, serendipity illustrates exactly what Walpole had in mind—discovering something valuable that he was not in quest of.

In a broader sense, serendipity includes luck in discovering the means to a goal one is in quest of. An example is Paul Ehrlich's discovery in 1909 of Salvarson, the most effective treatment for syphilis prior to penicillin.[6] In his early work, Ehrlich seeks general cures for bacterial and parasitic diseases, but he refocuses his research after coming to believe that some compound of arsenic might cure syphilis. Proceeding with exceptional patience on the basis of trial and error, he tests 605 compounds before discovering, by luck, that Salvarsan is effective. If we widen serendipity to include luck in the pursuit of a sought-after goal, as in this example, serendipity is ubiquitous in science.

Does the luck of Fleming and Ehrlich somehow diminish the scientific credit they deserve, given that it is something that happens to them rather than something they do? Surely not. Intelligence and commitment suffuse their inquiries, proximally and distally. Proximally, in the circumstances at the moment of discovery, they manifest "sagacity"—curiosity and imagination, alertness and perceptiveness, and creative insight in discerning the significance of the accidents that occur. Distally, in preparing to make discoveries, they study science with determination and perseverance, and they commit themselves to worthwhile lines of research. They are not "merely lucky," if that suggests their agency is unimportant. In general, scientists' luck emerges on a foundation of training and toil.[7] Hence Pasteur's observation: "chance favors only the prepared mind."[8]

Turning from scientific to moral credit, does Fleming and Ehrlich's luck reduce the moral recognition owed to them for saving lives and alleviating suffering? Again, surely not. Their moral contributions, like their scientific contributions, emerge from an inextricable mixture of agency and luck, commitment and happenstance. We credit them for making moral contributions through effort and skill, not for having good luck. More exactly, we credit them *for* contributions that are partly due to luck, but *because* of their agency, insight, skill, and perseverance.

At a personal level, overemphasis on luck raises hackles. Scientists take umbrage when their discoveries are portrayed as dumb luck, thereby diminishing the credit due them. As Robert K. Merton and Elinor Barber observe, "to be considered lucky is undesirable—it implies that achievements are really undeserved and that the lucky individual cannot be counted on to perform reliably. (If he is just lucky, after all, luck might easily desert him.)"[9] In contrast, serendipity, which combines good luck with sagacity, might be judged positively when the sagacity is highlighted and the luck downplayed, or judged negatively when the element of luck is emphasized.

More needs to be said, however, for the general role of luck in making assessments of moral responsibility has proven especially contentious among philosophers. That role is directly relevant to scientific serendipity and credit, even though it requires a brief digression into broader issues about agency and accountability.

Moral Luck and Merit

The serendipity involved in the moral contributions of Fleming and Ehrlich are instances of *moral luck*, that is, luck affecting conduct and character. Some philosophers, including Nicholas Rescher and Immanuel Kant, deny that moral luck exists. They view morality as solely about what we can control, whereas luck is something we cannot control. For them, "moral luck" is an oxymoron because it suggests that actions are simultaneously under our control and not under our control (in the same respects).

Moral contributions turn on actual outcomes, produced by a mixture of choice and chance. Is it perhaps more reasonable to praise Fleming and Ehrlich for their commitment and efforts, rather than for their actual achievements which, because of luck, they could not fully control? In general, should we praise and blame individuals for their efforts, rather than for their actual achievements involving luck?

Rescher thinks so: "What ultimately matters for the moral dimension is not achievement but endeavor," and "no matter how things eventuate, the goodness of the good act and the good person stands secure from the vagaries of outcomes."[10] We ought to praise people for their intentions and efforts, as well as for outcomes they can be expected to foresee and control, but not for successes which are due to luck.[11] Applying Rescher's view to our cases, apparently we should praise Fleming and Ehrlich for their effort and skilled work, but not for their actual contributions in saving human lives, given that those contributions involve the vagaries of chance outcomes. Presumably they should not be praised any more highly than numerous other forgotten scientists who worked with comparable diligence and skill, but who were not lucky in making momentous discoveries.

Although Rescher is exquisitely attuned to the role of luck in human affairs, he attempts to insulate morality as the one area impervious to luck. Ultimately there is "no such thing as *moral* luck."[12] Insofar as luck is involved, Rescher largely separates character and consequences. What matters morally is having a good character, with good intentions and competency, but not one's actual success.[13] Here he agrees with Kant: The morally good will "is not good because of what it effects or accomplishes, because of its fitness to attain some proposed end, but only because of its volition, that is, it is good in itself. . . . Usefulness or fruitlessness can neither add anything to this worth nor take anything away from it."[14]

This view captures one thread in ordinary moral understanding: we do esteem good motives and intentions, effort and skill. Yet, Rescher and Kant's view stands directly at odds with another thread: we also value good consequences and take them into account in evaluating persons and their actions, even when we know that luck plays a significant role. Saving millions of lives counts morally, and Fleming and Ehrlich deserve much credit for that outcome, despite the role of luck. In exclusively emphasizing motives and intentions, Rescher and Kant effectively sever the tie between actions and responsibility for outcomes. Actual outcomes are relevant to moral assessments of an individual because, as Thomas Nagel points out, "how things turn out determines what he has done," and "we judge people for what they actually do or fail to do."[15] More exactly, we judge people for their actions and for the consequences of their actions, as well as for their intentions and motives. To use Nagel's example, the actual outcome of the American Revolution established whether Washington, Jefferson, and Franklin are heroic founders of a great country or instead tragic failures who blame themselves on the way to the scaffold for worsening British oppression of their compatriots. As with most major historical events, the actual outcome involves luck, over which they had little control. Likewise, we justifiably praise Fleming and Ehrlich for their creative discoveries of life-saving drugs, not merely for their commitment, intelligence, and good intentions. Indeed, we cannot know what they discovered without knowing about the outcomes of their research.

Understandably, Rescher and Kant want to ensure that responsible persons receive credit for their good intentions, that they are not blamed for unforeseeable and unavoidable consequences of good-faith efforts to do what is right. They also want to ensure that wrongdoers are held responsible for their bad intentions and choices, even when by luck they avoid causing harm. These aims can be met, however, without denying the moral significance of luck in bringing about morally desirable results. We can give due credit for the sincere effort and hard work of those who are unlucky while still appreciating achievements involving a combination of effort and luck. In tune with ordinary moral understanding, individuals deserve credit for their actual contributions, as well as for their good intentions and sincere efforts, even when their contributions involve elements of luck.

I suspect that Rescher and Kant overemphasize intentions and motives because of their religious views. They understand the "true" moral self as a spiritual soul whose character is accurately discerned by a deity who judges us for our intentions rather than actual accomplishments. I also grant that religious perspectives might indeed transform ordinary views of what is praiseworthy. My concern, however, is with ordinary secular understandings of morality. Those understandings justify praise for achievements involving serendipity—the synergy of luck and sagacity. They also justify blame for wrongdoing in many instances where luck is involved.

Moral Skepticism and Blame

The deepest worry about moral luck is that it seems to make nonsense of blaming individuals for wrongdoing, including scientific misconduct (discussed more fully in chapter 6). It is unfair to blame people for their bad luck, something over which they have no control, and yet ordinary moral practices seem to do just that.[16] Ordinary judgments about moral responsibility assume that individuals are in control of what is being evaluated, that they have options among which they freely choose. But once we become attuned to how luck is ubiquitous in human affairs in ways that determine outcomes and options, we should radically revise ordinary morality, according to Thomas Nagel and Bernard Williams.

Nagel insists: "it seems irrational to take or dispense credit or blame for matters over which a person has no control, or for their influence on results over which he has partial control."[17] Williams agrees: "anything which is the product of happy or unhappy contingency is no proper object of moral assessment, and no proper determinant of it, either."[18] Virtually all actions involve some element of luck. And luck sometimes has a dramatic impact on moral matters. Thus, the difference between guilt for manslaughter and guilt for reckless driving can be a matter of luck, depending on whether a pedestrian happens to walk in the driver's path as he speeds through a red light.[19] And the difference between attempted and successful murder can depend on whether the shooter's bullet was deflected by a low-flying bird.[20]

Nagel and Williams believe these examples show that ordinary morality involves blaming people for their bad luck—specifically, blaming the driver for the presence of the pedestrian, and blaming the shooter for the absence of the victim's guardian bird. Such blame is unfair, and in that regard condemned by morality itself. Hence, they allege, there is a kind of incoherence in our ordinary ways of praising and blaming. Stated in another way, morality both accepts as fair and condemns as unfair the practice of blaming people for their bad luck. As such, ordinary morality is a muddle and needs radical revision.

I disagree. Morality does not require blaming victims of bad luck, and therefore it does not stand in need of overhaul. Quite simply, the driver is guilty of manslaughter because he culpably failed to drive responsibly, and the reckless driver is blameworthy for his reckless driving. Both can reasonably be expected to obey the law and to foresee the possibility of danger involved (if not all the specific dangers), and hence each is accountable for the harm caused. Again, the murderer is culpable because he intentionally set out to kill an innocent person and succeeded, whereas the attempted murderer commits a lesser, but still very serious, crime of attempted murder. The same applies to scientists whose misconduct leads to great harm or, through luck, minor consequences.

We hold people responsible for their misconduct but also for the actual harm their misconduct causes. More fully, we blame individuals *for* actual outcomes, which result from a nexus of choice and chance, but *because* of their conduct which is under their control. We do not blame them for or because of their bad luck. Luck plays a role in our moral assessments, but there is nothing unfair involved as long as there is ample room for agency to interact with luck in relevant ways.

As Margaret Urban Walker reminds us, morality is a response to the real world, and in that world there is a synergy of choice and chance, which can never fully be pried apart. This synergy is tacitly acknowledged by persons of integrity who can be trusted and depended upon to accept responsibility. It is also appreciated by creative scientists. Ordinary morality, by which I mean everyday moral understanding when it is reasonable, is based on the way the world is, and the world inextricably combines areas of control and chance. Assessments of scientists' moral contributions are based on the combination of sagacity and luck. Likewise, assessments of reckless drivers take into account the nexus consisting of what they can control (not driving recklessly) and the luck they cannot control (pedestrians appearing or not appearing at a given moment).

Walker says that "responsibilities outrun control" in ways that refute the idea that "people cannot be morally assessed for what is due to factors beyond their control."[21] It is more accurate to say, however, that people are responsible in light of what they can control, in situations involving a nexus of control and chance. Certainly, responsibilities do not outrun blame. For example, I am responsible for dealing reasonably with the medical problems of my young children, which means I have obligations in these areas. Because luck is involved in what those medical problems are, as well as in the outcomes of various treatment options, my responsibilities outrun my control. It does not follow that blame and credit outrun control. Having responsibilities for matters involving a nexus of choice and chance is one thing. Blameworthiness and praiseworthiness are further matters that involve assessments of both responsibilities and a complicated set of reasonable excuses for failing to meet responsibilities.

Here we need to keep distinct the several meanings and aspects of responsibility: having responsibilities (obligations), accountability for meeting responsibilities, praiseworthiness for good actions, the virtue of being responsible by regularly meeting obligations, liability for paying damages or compensation (for accidentally and blamelessly causing property damage), and blameworthiness for bad conduct.[22] Blame and credit pertain to how we deal with our responsibilities. They target our choices and areas of control, rather than our luck.

Life Plans

I have focused on praise and blame for particular actions during a career in science. Luck also plays a role in selecting careers in the first place. Williams is especially interested in the role luck plays in justifying life plans. He uses an example of an artist, a fictionalized version of Gauguin, which applies equally to scientists. Gauguin is morally anguished for abandoning his wife and children, leaving them in a grim situation, in order to pursue his painting in Tahiti. In Gauguin's eyes, the abandonment could be justified only if it enables him to become a successful painter—successful in terms of doing creative work, rather than obtaining fortune and fame. In the eyes of most observers, the decision to leave his family can be justified only if his painting provides a significant good for the world.[23] Such success is by no means assured. It is substantially a matter of luck, both the external luck of favorable circumstances and internal luck in having genuine talent as a painter. Justification of the abandonment can only be retrospective, in light of outcomes.

The example will not be convincing if we think that artistic and aesthetic contributions lack relevance to the moral issue of abandoning one's family. Even if we count those contributions as morally relevant, however, it is by no means obvious that success as a painter could justify Gauguin's actions—as opposed to excusing or mitigating his blameworthiness. Williams claims that "the *only* thing that will justify his [Gauguin's] choice will be success itself."[24] That seems too strong. For, we also need to consider such matters as Gauguin's vocation ("calling"), authenticity, meaningful life, and happiness. For example, if Gauguin were to die tragically after showing promising work, we might say he was authentic in pursuing his talent even though it did not work out. Still, Williams has a point. We do take into account outcomes in assessing decisions concerning careers.

Williams also indicates that the point could be made with other examples. We might think, for example, of Einstein's neglect of his family in the pursuit of his scientific interests. His marriage to Mileva Maric, who was a superb mathematician, contributed to his work in developing the special theory of relativity in 1905.[25] A decade later, however, when he developed the general theory of relativity, he had essentially abandoned Mileva and their two children to concentrate entirely on his research. He was not a good husband and father, although he made supportive gestures such as sending Mileva the money from his Nobel Prize (a more modest amount than today). He acknowledged that throughout his life he was devoted to science in a degree that rendered family and personal relationships entirely subordinate. If Einstein's career in physics had taken an unlucky turn, we might be less forgiving of how he treated his family.

To conclude, moral luck, both in circumstances and outcomes, affects all aspects of our lives. Specifically, it enters into our moral assessments of conduct, character, and careers. Acknowledging moral luck should not make us skeptical about morality or lead to radical revisions of it. It should, however, make us more cautious about blaming victims and bestowing credit unfairly.

In matters involving choice and chance, we should take care to target assessments on choices for which individuals are properly held accountable in light of their responsibilities. Let us next consider these assessments more fully as they concern scientific misconduct.

Notes

1. Horace Walpole, Letter to Horace Mann, 28 January 1754, in *The Yale Edition of Horace Walpole's Correspondence*, ed. W. S. Lewis (New Haven, CT: Yale University Press, 1937–1983), 20, 407–411. Serendip is today Sri Lanka.

2. Nicholas Rescher, *Luck* (Pittsburgh: University of Pittsburgh Press, 1995), 24. Luck does not mean the absence of causation.

3. Gwyn Macfarlane, *Alexander Fleming* (Cambridge, MA: Harvard University Press, 1984); Gilbert Shapiro, *A Skeleton in the Darkroom: Stories of Serendipity in Science* (San Francisco: Harper & Row, 1986), 41–58.

4. Eric Lax, *The Mold in Dr. Florey's Coat* (New York: Henry Holt and Company, 2005), 16–24.

5. On types of moral luck see Donna Dickenson, *Risk and Luck in Medical Ethics* (Cambridge: Polity Press, 2003), 4–7; and Daniel Statman, "Introduction" in *Moral Luck*, ed. Daniel Statman, (Albany: State University Press of New York, 1993), 11. I focus on outcome and circumstantial luck, setting aside luck in the formation of character—a topic connected with the thorny debate over free will.

6. James H. Austin, *Chase, Chance, and Creativity* (Cambridge, MA: MIT Press, 2003), 86–91.

7. Royston M. Roberts, *Serendipity: Accidental Discoveries in Science* (New York: John Wiley and Sons, 1989), 244–47.

8. Quoted by Dean Keith Simonton, *Creativity in Science* (New York: Cambridge University Press, 2004), 10.

9. Robert K. Merton and Elinor Barber, *The Travels and Adventures of Serendipity* (Princeton, NJ: Princeton University Press, 2004), 170–71.

10. Rescher, *Luck*, 155–56, 162.

11. Rescher, *Luck*, 159.

12. Rescher, *Luck*, 162, 158.

13. Rescher, *Luck*, 152.

14. Immanuel Kant, *Practical Philosophy*, trans. and ed. Mary J. Gregor (New York: Cambridge University Press, 1996), 50.

15. Thomas Nagel, *Mortal Questions* (New York: Cambridge University Press, 1979), 29–30, 34.

16. Samuel Butler satirizes such practices in *Erewhon* (New York: New American Library, 1960 [1872]).

17. Nagel, *Mortal Questions*, 28.

18. Bernard Williams, *Moral Luck* (New York: Cambridge University Press, 1993), 20.

19. Nagel, *Mortal Questions*, 25.

20. Nagel, *Mortal Questions*, 29.

21. Margaret Urban Walker, *Moral Contexts* (Lanham, MD: Rowman & Littlefield, 2003), 22, 26.

22. Claudia Card, *The Unnatural Lottery: Character and Moral Luck* (Philadelphia: Temple University Press, 1996), 28; Judith Andre, "Nagel, Williams, and Moral Luck," in *Moral Luck*, ed. Daniel Statman, 127.

23. Williams, *Moral Luck*, 37.

24. Williams, *Moral Luck*, 23, emphasis added.

25. Andrea Gabor, Einstein's Wife (New York: Viking, 1995), 3–32; Abraham Pais, *Subtle Is the Lord: The Science and Life of Albert Einstein* (Oxford: Oxford University Press, 1982).

CHAPTER SIX

Scientific Misconduct

Creativity has a dark side. Like any great passion, the ambition to be creative can become excessive and unbalanced, overshadowing respect for truth, other professional responsibilities, and even one's own creative aims. Scientific misconduct betrays moral ideals, tarnishes the public image of science, and impugns the reputation of universities and corporations. It can even damage entire nations, as when South Korean Hwang Woo Suk was exposed in 2005 for faking data in experiments on cloning. It is not surprising, then, that scientific misconduct is currently a serious concern, in addition to being a familiar theme in literary works such as Aeschylus's *Prometheus Bound*, Mary Shelley's *Frankenstein*, Robert Louis Stevenson's *Dr. Jekyll and Mr. Hyde*, and Goethe's *Faust*. I begin with dishonesty and then comment on whistleblowing in response to scientific misconduct. Next I turn to failure to respect experimental subjects. I conclude with Faustian bargains that widen the topic of scientific misconduct to include the overlap of personal and professional life.

Honesty and Honors

Prometheus uses deception and cunning to steal fire from Zeus. In one version of the legend, Prometheus is a small-time crook who violates the gods' trust and thereby causes human misery. In another version, he is a heroic benefactor whose skillful ruse greatly advances human progress. Either way, by linking creativity and deception the legend foreshadows contemporary concerns about scientific integrity.

The extent of scientific misconduct is unclear, and there are disagreements about how broad the definition of scientific misconduct should be. Not long ago it was commonplace to downplay scientific misconduct as the province of a few deranged individuals. That perception is changing. In a 2005 study, one-third of scientists admitted to some form of scientific misconduct in the past three years: "less than 1.5% admitted to falsification [of data] or plagiarism. . . . But 15.5% said they had changed the design, methodology or results of a study in response to pressure from a funding source; 12.5% admitted overlooking others' use of flawed data; and 7.6% said they had circumvented minor aspects of requirements regarding the use of human subjects."[1]

To begin with an extreme case, Mark Spector struck his supervisors at Cornell University as a prodigy because he managed to purify several new enzymes that render cancer cells less efficient in their functioning.[2] To all appearances, he relied on standard experimental techniques in using radioactive phosphorus to trace the manufacture of proteins in the body. His discoveries came at an astonishing rate and suggested a revolutionary explanation of the genesis of cancer. Researchers around the world, however, were unable to duplicate his experiments. Another Cornell researcher identified the fraud: Spector was using radioactive iodine, not just phosphorus, to fabricate the data he then claimed to discover. He engaged in additional fraud, including lying about an undergraduate degree he never earned. Spector is atypical; he was more con artist than creative scientist, and his motives were to gain recognition by merely appearing to be creative.

More often, scientific fraud involves ambitions to be creative, as well as appear as such, and scientists become convinced that their work is moving toward significant discoveries. William Summerlin was a competent researcher who coauthored a number of papers with Robert Good, a famous immunologist.[3] Summerlin believed he was close to a breakthrough in transplanting human tissue without the usual problems of tissue rejection. He was also convinced he had successfully used the technique on mice, and that additional grants would generate important results. When he outlined his research at the meetings of the American Cancer Society, the media eagerly publicized it. But other scientists were unable to replicate his results. In particular, Robert Good was unable to confirm Sommerlin's recent experiments, and he was about to publish an article that would undermine his future funding. On the morning of a crucial meeting with Good, Sommerlin used a felt-tip marker to blacken the patches on two white mice, creating an appearance of successful black-mice to white-mice grafts. Good was conned, but later in the day a lab assistant who noticed the marker ink was running blew the whistle on him.

Sommerlin's motive was not wealth. Instead, it was the desire to gain funding for research he found promising, as indeed it might well have been. One scientist who reviewed Sommerlin's research speculated that he had made innovative transplants early in his career but was unable to repeat them: "Being absolutely convinced in his own mind that he was telling a true story, he thereupon resorted, disastrously, to deception."[4] In this way, Sommerlin illustrates how the desire to be creative can both drive and distort scientific inquiry.

Deliberate misconduct is only part of the problem. Self-deception and gullibility probably play a larger role.[5] Researchers who want to make creative discoveries can be all too eager to see what they want to see in order to establish their discoveries. The result can be bias and sloppiness, which in science are serious failings located somewhere between cynical fraud and excusable error.

Cold fusion is a famous example. Stanley Pons, chair of the University of Utah's chemistry department, and his collaborator Martin Fleischman of Southhampton University, were convinced they had discovered a potentially limitless source of energy using a simple electrochemical reaction that caused hydrogen nuclei to fuse, producing excess heat and radiation.[6] Rather than wait until their results were published and subjected to additional testing, they announced their discovery at a press conference hastily called by the president of the university. Because Pons and Fleishchman were well-respected electrochemists, their announcement sparked an explosion of research, at great expense, most of which failed to confirm their results. Pons and Fleishchman were not guilty of fraud, nor were they routine victims of error. They were their own dupes.

Self-deception takes many forms, ranging from motivated false beliefs to willful ignorance and purposeful evasion of evidence. Ironically, self-deception occasionally contributes to creativity by bolstering self-confidence in pursuing promising lines of research. In this way, self-deception is sometimes truth-promoting when it remains at the level of motivation rather than corroboration.[7] But, as in the Pons-Fleishchman case, self-deception and bias are more often unreliable cognitive practices that lead to violations of scientific standards. In science it is not enough to avoid fraud. What is needed, as Richard Feynman reminds us, is "a kind of utter honesty—a kind of leaning over backwards." Thus, in reporting experiments scientists should lean over backwards to identify alternative explanations for the results and also provide all information that can aid other researchers.[8] Only rigorous truthfulness ensures trustworthiness, so that the results of experiments can reliably enter into future research by other scientists.

Fully explaining why scientists sometimes fail to lean over backwards to be honest is a task for psychologists and social scientists. But our examples suggest a major role for desires to be creative and to appear as such. In addition, more general concerns for reputation (fame, recognition) and money are interwoven with desires to be creative. These general human motives, however, should not be equated with the specific motives for much scientific misconduct. As we noted in chapter 3, Alasdair MacIntyre attempts to explain professional misconduct in terms of generic motives such as money, power, and recognition, which override "the ideals and the creativity of the practice."[9] This human-nature explanation of wrongdoing fails to pinpoint that what is at stake is reputation *as* a scientist, or *as* a biologist or physicist. Furthermore, in most instances of scientific misconduct there is a desire to be a creative scientist, as well as to appear to be such.

Robert K. Merton provides a necessary emendation to MacIntyre's explanation of wrongdoing. In drawing attention to the importance of originality as an institutional norm in science, Merton highlights how that norm generates powerful incentives. Scientists are constantly reminded that the highest excellence consists in creative advancement of their disciplines. Recognition for being creative, then, amounts to recognition for doing one's work with excellence. That recognition, in turn, becomes the primary intellectual property of the scientist. That is, because the content of scientific studies becomes public property upon publication, ownership in science consists largely of the acknowledgment by others of one's creative outcomes.[10] Merton concludes that much selfish motivation is simply the psychological counterpart of institutionalized scientific norms. When this emphasis becomes extreme, the norm of creativity can contribute to misconduct: "Competition in the realm of science, intensified by the great emphasis on original and significant discoveries, may occasionally generate incentives for eclipsing rivals by illicit or dubious means."[11]

Applying the definition of creativity as purposefully bringing about something new and valuable, we can say that Merton accents *the new*—being first in making an original discovery. In contrast, David L. Hull accents *the valuable*—making a significant contribution. He notes that "the most important recognition that scientists receive for their work is its use by scientists working in the same area, including of course their closest competitors. No scientist can get very far without using the work of his or her fellow scientists and in using it tacitly admits its value."[12] Once the value of scientists' ideas is established within their domain of research, they might also desire wider public honors. But recognition depends on the perception within their field that they have made a significant discovery—epistemically significant in advancing further research, and perhaps technologically and morally significant as well. Wrongdoing can occur when this aspiration for recognition becomes excessive. "Excessive," in this context, does not refer to felt intensity or motivational strength, but rather to the neglect of other important moral values that should balance and limit the desire for recognition.

Whistleblowing

Scientific misconduct usually remains hidden until whistleblowers reveal it. Historically, the scientific community assumed there is little need for whistleblowing. Science is a self-correcting enterprise whose routine procedures correct both error and fraud. Experiments, especially important ones, are subjected to corroboration by retesting, which usually uncovers error before it has a chance to do much harm. This is true whether a scientist commits fraud in reporting data as in the Sommerlin case, or instead is self-deceived or sloppy as in the Pons and Fleishchman case. This optimistic view of science overlooks that only a small percentage of experiments are repeated, given the cost of research and the difficulty of getting such routine experiments published in journals. Moreover, scientists need to trust each other's work, and fraud can misdirect future research and waste important resources.

Whistleblowers, then, play an important role in disclosing dishonesty. Such disclosures are rarely as straightforward as a lab assistant noticing marker ink smearing on a white rat. Coworkers who suspect that bad science has occurred often do not have a knock-down proof of fraud. Furthermore, allegations of sloppy or dishonest science provoke countercharges of disloyalty. Even the most responsible and courageous whistleblowers usually suffer unhappy fates, often by being fired and prevented from finding comparable work.

Whistleblowing is usually defined as conveying information about a serious moral problem in one's organization outside approved organizational channels to someone in a position to take action on the problem, either inside or outside the organization. These days nearly all organizations claim to have open-door policies that celebrate free communication within the organization. In practice, however, there are strong pressures against challenging individuals with authority or in other ways being perceived as a troublemaker. Hence we need to add to the definition of whistleblowing that it might consist in communicating information within approved organizational channels when doing so goes against organizational pressures, such as the wishes of supervisors and colleagues.[13]

A large literature is devoted to issues about when whistleblowing is justified. Usually it is argued that whistleblowing falls under the general obligation to prevent harm to others, whether that obligation is grounded in utilitarian ethics (promote the most good for the most people), right ethics (respect human rights), or virtue-oriented ideals of decency and justice. Some philosophers attempt to state rules about how to proceed, for example, by first working within approved organizational channels and acquiring evidence about serious wrongdoing.[14] Other philosophers see whistleblowing situations as too complex for simple rules and instead identify factors that should be considered and weighed.[15] Alternatively, some ethicists focus more on personal integrity than on the duty to prevent harm to others. Members of organizations who observe serious misconduct become implicated in the wrongdoing, coconspirators as it were, unless they engage in whistleblowing.[16]

In my view, preventing harm and maintaining one's integrity are complementary reasons that enter into most whistleblowing cases. Moreover, in science an additional character-related consideration is at stake. Science depends on the shared commitment of all participants to the pursuit of truth. Upon learning of serious misconduct in reporting research, scientists have special responsibilities to take appropriate steps to act on that ideal.

The duty to whistleblow on scientific misconduct is prima facie, that is, it has limits and exceptions, and it needs to be balanced with other moral duties. Specifically, potential whistleblowers need to take into account their obligations to employers and colleagues. In many cases, the best way to balance these dual obligations is to have procedures in place within organizations that allow whistleblowers to request that research be reviewed and if necessary repeated. In practice, however, whistleblowers are rarely allowed to proceed in a measured manner, and it might take a novelist to delineate the complexities that can arise.

Allegra Goodman is such a novelist. In *Intuition*, she explores the moral ambiguities surrounding creative ambitions and whistleblowing alike. Cliff, a young scientist, becomes convinced he has hit upon a potentially new cancer treatment. One codirector of the lab where he works urges quick publication of his results to avoid being scooped by competing labs, while the other codirector is more worried about the embarrassment of acting prematurely. Robin, Cliff's coworker and girlfriend, is charged with confirming his results. As part of her research she studies Cliff's lab books and finds discrepancies between them and the data he reports in the published version of the experiments. Robin is frustrated in her career and jealous of Cliff's successes. She is also convinced that Cliff's publication suppresses relevant data, exaggerates other data, and possibly collates information from different experiments as if they were one. When she conveys these beliefs to the lab directors, she is met with extreme hostility by one director and other coworkers. Eventually she expresses her view to a colleague outside the lab, at which point events move beyond her control. Indirectly, she precipitates a federal investigation that initially criticizes Cliff as a fraud and subsequently portrays him as sloppy. Additional experimentation by other labs fails to confirm the results of his research. In the end, Cliff's conduct remains open to different interpretations. In many ways, Robin acted admirably, and yet perhaps a broader sense of tolerance for ambiguity might have made her more cautious in how she proceeded.[17]

Goodman's novel is loosely based on the real and infamous Baltimore case, named after David Baltimore, a Nobel Laureate who later became president of two prestigious universities.[18] In 1986 Baltimore headed the large MIT-affiliated Whitehead Institute and coauthored a paper with Thereza Imanishi-Kari (among others), who directed a smaller lab at MIT. The paper, which was published in the journal *Cell*, reported mice experiments on what the researchers believed was an important issue in human immunology. Imanishi-Kari assigned Margot O'Toole, a postdoctoral fellow, to duplicate her experiments while the essay was under submission. Imanishi-Kari, who was known for her volatile personality, put enormous pressure on O'Toole to get quick results, but O'Toole was unable to confirm her supervisor's data, despite working with great dedication and skill for much of a year.

In attempting to understand why her experiments were not working, O'-Toole came across seventeen pages in lab notebooks kept before she was hired that revealed other researchers, perhaps including Imanishi-Kari, had similar difficulties with their experiments. The pages seemed to indicate that some of the data reported in the journal article was inaccurate and not properly reported. At some point O'Toole requested that the essay be recalled and the data corrected. Such a request is entirely in keeping with the highest standards of scientific integrity. But Imanishi-Kari and Baltimore refused to comply. O'Toole then reported the event to colleagues at another school. At first Imanishi-Kari was found guilty of misconduct, but an appeal board concluded there was insufficient evidence for that finding.

Questions lingered. Tests by the Secret Service established improprieties in Imanishi-Kari's lab books. At the very least she had engaged in sloppy science, and it seems she was also dishonest about one of her degrees. As for Baltimore, he was mostly combative and uncooperative throughout the controversy. He should have agreed that the essay not be published until further experiments were performed, especially since he had not fully participated in overseeing the experiment done in Imanishi-Kari's lab. Instead he dug in, convinced that nothing was amiss and manifesting what many considered arrogant pride. As a result, what should have been a minor scientific conflict escalated into a national controversy within the scientific community and involved federal hearings and investigations.

Both the Baltimore case and Goodman's case illustrate how scientific creativity can involve large egos and unhappy collaborations in ways that complicate the ethics of whistleblowing. Whistleblowers are often junior scientists or even lab technicians at the lower end of the hierarchy of power. Many whistleblowers, like O'Toole, are enormously courageous individuals who uphold the highest professional standards. To be sure, not all whistleblowers act responsibly and prudently, and at least half of them act from dubious motives and evidence, thereby unfairly damaging the reputations of their organizations. In light of the Baltimore case, universities and corporations have developed new procedures to encourage responsible reporting of scientific conduct without unduly harming other scientists and organizations. Much more needs to be done, however, and all of us share a responsibility to ensure that responsible whistleblowers are protected.

Truth Seeking and Respect for Persons

So far we have focused on duplicity that undermines the commitment to truth at the heart of science. Even this commitment to truth, however, can become excessive and undermine other important values, in particular respect for experimental subjects. Research on humans depends on the willingness of experimental subjects to take risks. That willingness must be based on informed, voluntary consent, which in turn is grounded in respect for the dignity and worth of individuals.

Josef Mengele is a case of monstrous immorality—sheer evil. In addition to being a decorated officer and a Nazi ideologue, Mengele was a reputable physician-researcher who in 1943 asked to go to Auschwitz because of the research opportunities it afforded. He was fascinated with twins and used them to sort out biological and environmental causes of human features, thus saving dozens of twins from execution. He also willingly agreed to sort persons as they came off the trains at Auschwitz, sending some individuals to the barracks and others to instant death. In conducting his experiments, Mengele mixed seemingly friendly and paternal kindness toward the twins with demonic cruelty, including murder of some twins to perform autopsies for experimental purposes. After the war, two twins described him as "like Dr. Jekyll and Mr. Hyde."[19]

How should we understand Mengele's split between truth seeker and tyrant? In Stevenson's *Dr. Jekyll and Mr. Hyde*, a psychological split is brought about by Dr. Jekyll in experiments on himself. He discovers a drug that enables him at will to transform his Victorian character into an alter ego, the callous and cruel Mr. Hyde. Rather than resolving his inner divisions, the two selves grow increasingly at odds, until Hyde kills Jekyll, and hence himself. In fact, Hyde is not entirely evil, and he manifests a vitality missing in Jekyll.[20] Psychiatrist Robert Jay Lifton understands Mengele and other Nazi doctors in terms of an analogous psychic doubling: "the division of the self into two functioning wholes, so that a part-self acts as an entire self."[21] Psychic doubling is an unconscious process which facilitates evasion of guilt. I do not question that such doubling occurs, but I doubt it explains Mengele. Rather than having an unconscious inner schism, Mengele was more like a unified person acting consistently according to a racist ideology that glorified Aryans and dehumanized Jews and other minorities.[22]

Construed less literally, the notion of divided personalities becomes a metaphor for the kinds of conflicting motivations that can lead to wrongdoing. For example, long after the Nuremberg Code mandated the informed and voluntary consent in human experimentation, the infamous Tuskegee Syphilis Study continued.[23] The study sought valuable knowledge about the natural course of untreated syphilis, but it did so by telling poor and poorly educated black males with syphilis that they would be treated for their "bad blood," when in fact many of them were given placebos. At the time a partially effective therapy, Salversan, was available but withheld because of lack of therapeutic funding, and years before the study was ended penicillin became available as an effective treatment. At least twenty-eight males died needlessly during the four-decade study.

The Tuskegee study was funded in part by the United States Public Health Service, with the involvement of a surgeon general who was a member of the American Eugenics Society. The U.S. government had a hand in much additional scientific misconduct, especially where national security interests played a role.[24] For example, numerous radiation experiments were conducted without informed consent on the part of research subjects or bystanders, such as those in Utah and Nevada that exposed thousands of people to the toxic clouds from nuclear explosions. In these instances, patriotism adds to the usual motives for creative activity, recognition, and employment.

Tuskegee and Mengele remain cautionary tales about how the pursuit of truth is always intertwined with motives that range from personal recognition to money. More recently, researchers' temptations have become more nuanced but they remain ubiquitous. Consider, for example, conflicts of interest centered on money. Conflicts of interest are situations in which researchers' professional judgment is compromised, or in which there is even the appearance of such compromise. An example is the 1998 case of Jesse Gelsinger, who was seventeen and had a long history of medical complications from a rare genetic disorder.[25] He and his father gave consent to participate in a University of Pennsylvania inquiry to determine acceptable levels of tolerance (for toxicity) of an experimental gene therapy. The protocol called for six groups of patients who would be given increasingly strong infusions of a genetically modified adenovirus, and Jesse was assigned to the highest level. Within four days he was dead from complications related to the injected material. Subsequently the father learned that the university research institute and its head were being funded by the drug company who owned the experimental material. (The institute director sold his share in the company for $13.5 million.) Questions were raised about whether informed consent had been obtained, and specifically about whether the risks had been fully divulged.

Conflicts of interest are proliferating as higher education becomes increasingly commercialized. Money-strapped universities eagerly enter into mutually profitable partnerships with corporations, especially biotech firms. Some of these arrangements are not in the public interest. For example, researchers at one university discovered a gene that causes breast cancer.[26] Although their experiments were funded by millions of taxpayer dollars, the researchers immediately patented the gene and gave monopoly rights to a company which severely limited the use of the gene by other researchers. When conflicts of interest are openly acknowledged, most university-corporation partnerships have more beneficial results. Yet too often they are not acknowledged. A 2000 study found only 3 of 250 research institutions required researchers to inform research subjects of financial conflicts of interest.[27]

Moral Ambiguity and Faustian Bargains

Dishonesty, whistleblowing, and informed consent are central topics in professional ethics, but Faustian bargains are perhaps less often discussed, especially when they center on the overlap of private and professional life. According to Goethe's version of the legend, Faust makes a morally ambiguous bargain. In return for a lifetime of striving after knowledge and experience otherwise foreclosed to him, he agrees to let the devil claim his soul upon death. The bargain causes much harm, including several deaths, yet his creative striving is sufficiently admirable that God decides to save him. The story is layered with meanings, and as a result the term "Faustian bargain" can refer to many things: (a) to take large personal risks (e.g., damnation) in pursuing a greater social good (e.g., new knowledge); (b) to pursue a personal good (e.g., creativity) by harming others; (c) to risk or sacrifice a personal good (e.g., one's soul) for another personal good (e.g., fame); (d) to undertake, as individuals or groups, projects that are ambiguous mixtures of good and bad. I will concentrate on the first meaning, taking risks in the hope of achieving a public good.

The American Association for the Advancement of Science's Committee on Scientific Freedom and Responsibility affirms a basic responsibility of scientists to use their knowledge to improve the quality of life. Because of their expertise and positions of authority, scientists are often best qualified to foresee dangers. To be sure, they are not required to abandon their careers in research to become public educators, and they are also bound by confidentiality agreements that place restrictions on warning the public of dangers. Nevertheless, as Kristin Shrader-Frechette points out, scientists have "a duty to accomplish good if it can be done at no great cost or sacrifice."[28] This general duty places special responsibilities on them to provide the public with scientific information relevant to the public good, as well as to avoid engaging scientific research on behalf of immoral causes. Exactly what these responsibilities amount to requires a case-by-case analysis.

The case of Fritz Haber combines the exemplary with the execrable. Haber received the Nobel Prize for discovering how to transform nitrogen from the atmosphere into ammonia, which in turn is used to make fertilizer. This nitrogen fixation process, which is still used today, vastly increases world food production. It makes possible the survival of two billion people, roughly one-third of the earth's population. But Haber also applied the process in military service to Germany during World War I. Initially he made nitric acid for conventional weapons, but later he orchestrated the development of poison gas. His motives combined scientific curiosity with fervent nationalism. He proudly wore a military uniform and oversaw the killing of hundreds of enemy combatants. Two decades later the country he served as a superpatriot turned against him, expelling him as a Jew. In a final irony, one of the poison gases he developed, Zyklon, was developed into the Zyklon B used in the genocide at Auschwitz and Treblinka, where some of his relatives were killed.

Daniel Charles points out that Haber confronted issues we still face today: "Haber lived the life of a modern Faust, willing to serve any master who could further his passion for knowledge and progress. He was not an evil man. His defining traits—loyalty, intelligence, generosity, industry, and creativity—are as prized today as they were during his lifetime. His goals were conventional ones: to solve problems, prosper, and serve his country."[29] Although Haber was tragically misguided, he did not rationalize away his responsibilities by claiming he was merely a morally neutral servant of his science and his country. Such rationalization is common today whenever scientists and engineers narrowly compartmentalize their responsibilities.

Compartmentalizing, in a neutral sense, is focusing one's efforts in one area while ignoring distracting factors. As such, compartmentalizing promotes creativity, given the absorption needed to solve demanding technical problems. "Compartmentalizing" takes on pejorative connotations, however, when it suggests excessive, unreasonable, immoral disregard for other responsibilities in pursuing narrow aims. The immorality might involve emotional absorption at the expense of balanced judgment, emotional distancing from research subjects or from people affected by one's research, and denial of any share of responsibility for the harm caused by one's research or technological work. Exactly when these things are morally objectionable, of course, depends on moral judgments about the responsibilities of scientists and engineers.

Both the extent and proper balancing of role responsibilities is much contested. Some scientists and engineers define their responsibilities narrowly, as performing their assigned tasks within universities or corporations, thereby serving the public good indirectly. In contrast, others view the responsibilities to the public as bearing on all foreseeable or likely uses of knowledge and technologies. Otherwise, no one ends up accepting responsibilities, given the sharp delineation of roles within modern corporations and other organizations. As an example, Arnold Pacey recalls the development of napalm explosive during the Vietnam War. Initially, victims could scrape it off their skin, thereby inhibiting its full effect. Chemists were asked to develop ways to make it stick, which they did by adding polystyrene. "For the chemists, this was a narrowly technical question, and they did not have to dwell on the way napalm would now 'keep on burning right down to the bone'. . . . Nobody came face to face with the reality of what they were doing."[30]

Finally, the most comforting form of compartmentalizing is to define one's role as a creative researcher, and to regard technological applications as entirely others' responsibility. This stance is often accompanied by the view that all knowledge should be freely open to investigation. There should be no forbidden knowledge. Let us next ask whether that is true.

Notes

1. Meredith Wadman, "One in Three Scientists Confesses to Having Sinned," *Nature* 435 (June 9, 2004), 718–19. The study was conducted by Brian Martinson of

the HealthPartners Research Foundation in Minneapolis, Minnesota. I should note that modifying the design of an experiment under pressure from a funding agency is not always a sign of misconduct.

2. William Broad and Nicholas Wade, *Betrayers of the Truth* (New York: Simon & Schuster, 1982), 63–73.

3. Broad and Wade, *Betrayers of the Truth*, 153–57.

4. Peter Medawar, quoted in Broad and Wade, *Betrayers of the Truth*, 156.

5. Broad and Wade, *Betrayers of the Truth*, 107–25.

6. John R. Huizenga, *Cold Fusion: The Scientific Fiasco of the Century* (New York: Oxford University Press, 1993); Gary Taubes, *Bad Science: The Short Life and Weird Times of Cold Fusion* (New York: Random House, 1993); Bart Simon, *Undead Science: Science Studies and the Afterlife of Cold Fusion* (New Brunswick, NJ: Rutgers University Press, 2002).

7. Mike W. Martin, *Self-Deception and Morality* (Lawrence: University Press of Kansas, 1986), 126.

8. Richard P. Feynman, *Surely You're Joking, Mr. Feynman!* (New York: Bantam, 1986), 311.

9. Alasdair MacIntyre, *After Virtue*, 2d ed. (Notre Dame, IN: University of Notre Dame Press, 1984), 194.

10. Robert K. Merton, "Priorities in Scientific Discovery," in Robert K. Merton, *The Sociology of Science*, ed. Norman W. Storer (Chicago: University of Chicago Press, 1973), 294–95.

11. Merton, *The Sociology of Science*, 311.

12. David L. Hull, *Science as a Process* (Chicago: University of Chicago Press, 1988), 309–10.

13. Mike W. Martin, *Meaningful Work: Rethinking Professional Ethics* (New York: Oxford University Press, 2000), 139; Mike W. Martin and Roland Schinzinger, *Ethics in Engineering*, 4th ed. (Boston: McGraw-Hill, 2005), 172–73.

14. Richard T. DeGeorge, "Ethical Responsibilities of Engineers in Large Organizations, *Business and Professional Ethics Journal* 1 (fall 1981): 1–14.

15. Gene G. James, "Whistle Blowing: Its Moral Justification," in *Business Ethics*, 4th ed., ed. W. Michael Hoffman, Robert E. Frederick, and Mark S. Schwartz (Boston: McGraw Hill, 2001), 291–302.

16. Michael Davis, "Some Paradoxes of Whistleblowing," *Business and Professional Ethics Journal* 15 (Spring 1996), 3–19.

17. Allegra Goodman, *Intuition* (New York: Dial Press, 2006), 313.

18. Horace Freeland Judson, *The Great Betrayal: Fraud in Science* (Orlando, FL: Harcourt, 2004), 191–243.

19. Quoted by Robert Jay Lifton in *The Nazi Doctors: Medical Killing and the Psychology of Genocide* (New York: Basic Books, 1986), 355.

20. Robert Louis Stevenson, *Dr. Jekyll and Mr. Hyde* (New York: Bantam, 1981), 100–101.

21. Lifton, *The Nazi Doctors*, 418.

22. Cf. Berel Lang, *Act and Idea in the Nazi Genocide* (Chicago: University of Chicago Press, 1990), 48–56.

23. James H. Jones, *Bad Blood: The Tuskegee Syphilis Experiment* (New York: Free Press, 1993).

24. Andrew Goliszek, *In the Name of Science* (New York: St. Martin's Press, 2003).

25. Ronald Munson, *Intervention and Reflection*, 7th ed. (Belmont, CA: Wadsworth, 2004), 3–8.

26. Jennifer Washburn, *University, Inc.: The Corporate Corruption of American Higher Education* (New York: Basic Books, 2005), xi.

27. Derek Bok, *Universities in the Marketplace: The Commercialization of Higher Education* (Princeton, NJ: Princeton University Press, 2003), 70.

28. Kristin Shrader-Frechette, *Ethics of Scientific Research* (Lanham, MD: Rowman & Littlefield, 1994), 67.

29. Daniel Charles, *Master Mind: The Rise and Fall of Fritz Haber, the Nobel Laureate Who Launched the Age of Chemical Warfare* (New York: HarperCollins, 2005), xvii. See also Sharon Bertsch McGrayne, *Prometheans in the Lab: Chemistry and the Making of the Modern World* (New York: McGraw-Hill, 2001), 58–78.

30. Arnold Pacey, *Meaning in Technology* (Cambridge, MA: MIT Press, 1999), 176–77. Pacey takes the example from Noam Chomsky, *The Backroom Boys* (London: Fontana/Collins, 1973), 23.

CHAPTER SEVEN

Forbidden Knowledge

God forbids Adam and Eve to eat from the Tree of the Knowledge of Good and Evil. When they disobey, He banishes them from the Garden of Eden to confront all the horror, but also the beauty and goodness, of this world. Conservative religious thinkers invoke this idea of forbidden knowledge, along with the notion of "playing God," to criticize scientific research they find objectionable. Yet even when secularized, the expression "forbidden knowledge" is resonant with danger and discovery, constraint and creativity. This secularized meaning shifts away from divine and omniscient authority toward democratic and fallible authority in overseeing scientific inquiry, whether through law and regulation, decisions about funding research, or self-regulation within the scientific community.[1]

Are there moral limits to what we should know, as individuals, societies, and humanity? If so, are there areas where scientific creativity should be curtailed? Unquestionably, there are sometimes good reasons to forbid the acquisition of, access to, and application of particular knowledge. There are also areas where the mere possession of knowledge undermines, at least for a while, the meaning an individual or an entire society finds in life. Nevertheless, in my view scientific inquiry should never be forbidden on the grounds that it threatens conventional beliefs and sources of meaning. At the same time, it is important to be clear about the justification for any restraints and restrictions, and most of this chapter is an attempt at clarification.

Reasons for Forbidding Knowledge

Reasonably or not, knowledge and inquiry might be morally forbidden on four types of grounds: acquisition, access, application, and mere possession. *Acquisition* of knowledge, to begin with, is immoral when it employs procedures that violate moral norms such as rights to privacy, informed consent, and safety. Egregious examples, discussed in chapter 6, include the Nazis experiments on Jews, Tuskegee Institute's syphilis studies, and the U.S. government's radiation research. Since the Nuremburg Code was written following World War II, a broad consensus has emerged concerning the moral principles governing research procedures, especially the duty to obtain informed consent of experimental subjects. To be sure, areas of controversy remain. Some ethicists, for example, condemn all experimentation on prisoners, given the coercive conditions under which they live, while other ethicists believe an absolute ban violates prisoners' rights.

Access to knowledge, which includes dissemination of information, sometimes ought to be forbidden to individuals and groups who are likely to misuse it. Some knowledge is best kept secret from the public, for example medical records, social security numbers, and some national security information. Again, it is desirable to withhold information from terrorists about how to manufacture chemical and biological weapons, although the Internet has made many such restrictions unenforceable.

Application of knowledge refers to technological uses of scientific and engineering understanding. Here forbidden knowledge includes how to build unacceptably dangerous technologies, such as compact nuclear bombs that a terrorist could carry in a suitcase. When inquiries seem likely to lead to extremely dangerous knowledge, and when the danger cannot be counteracted at reasonable cost, there is a clear rationale for restricting research. And when disagreements arise about which risks are acceptable, we must rely on the guidance of committees in making technology assessments. Even if the facts about dangers are clear, value differences can polarize opinions. Is destroying a human embryo an acceptable research technique, especially when the embryo is obtained with permission from fertility clinics and it would be destroyed anyway, or does that amount to killing a human being? Regulatory committees need to weigh and integrate relevant facts and values, working out reasonable compromises.

Most of us agree there are moral limits on how knowledge is acquired, disseminated, and applied, but what about the simple *possession* of knowledge? This question cuts to the heart of whether scientific creativity has any fixed limits. The scientific ethos, which emerged during the Enlightenment, values all knowledge. As voiced by Carl Sagan, the ethos insists "there are no forbidden questions in science, no matters too sensitive or delicate to be probed."[2] All truth, knowledge, and creative discovery are valuable, from the broadest explanatory principles to details about nature's working. Any justifiably forbidden knowledge consists solely in how information is pursued, disseminated, and applied, and even then prohibitions should be limited to situations of great dangers that we are not yet able to manage.

The scientific ethos is demanding and difficult to live with. Witness the long history of clashes between science and dominant religions. Copernicus's heliocentric view of the galaxy deeply threatened Catholic dogma. Darwin's theory of evolution profoundly challenged the religious beliefs of Victorians, and many fundamentalists today. And Freud and Kinsey, together with birth control pills, overthrew puritanical views of sexuality. Despite the controversies it generates, the scientific ethos has brought enormous good, and in some version it deserves our strong allegiance.

Which version? Contrary to Sagan, not all knowledge is inherently valuable. Much of it is utterly trivial, for example, knowing how many blades of grass are in my lawn (with no further research aim in mind). Moreover, scientific advances are Janus-faced in that the same knowledge or technology can have both good and bad uses. Thus, knowledge of nuclear reactions can be used to create electricity or make bombs, and Haber's nitrogen fixation process can be used to produce fertilizer or wage chemical warfare. This does not mean that knowledge is valuable only insofar as it has practical benefits. To affirm the scientific ethos is to value significant truth even when it is double-edged in its applications. Likewise, the scientific ethos celebrates the intellectual virtues, especially love of truth and commitment to creative discovery. The interesting question is whether the mere possession of knowledge should ever be forbidden because it threatens moral, religious, or political values. The scientific ethos says no, but that ethos has been challenged by some incisive philosophers of science, most notably Nicholas Rescher and Philip Kitcher.

Rescher: Revise the Scientific Ethos?

Let us begin with Rescher. In an insightful but confusing essay, "Forbidden Knowledge: Moral Limits of Scientific Research," Rescher asks whether the mere possession of knowledge should ever be forbidden. At first glance, he seems to answer that it should not. Without saying that all knowledge is inherently valuable, he contends that no knowledge is inherently immoral, such that to possess it "is morally inappropriate *per se*."[3] Moreover, there is a strong presumption in favor of pursuing truth wherever it leads and protecting that pursuit against such forces as superstition, bigotry, and intellectual laziness. Scientific inquiry should be constrained only when there are "clear and present dangers" in the areas of acquisition, access, or application of knowledge.[4]

This view seems to me correct. Rescher's essay is puzzling, however, because in discussing examples he seems go against his own thesis. In those discussions he seems to object to the mere possession of knowledge when it seriously deflates personal meaning and undermines valuable ways of life. Consider three illustrations. First, he says that research into the intelligence and abilities of ethical groups is immoral when it threatens democratic values about equal respect and opportunity.[5] He claims that this case pertains to the distribution of knowledge, that is, to the bad consequences of allowing access to this information. Yet, the case is apparently an instance where merely possessing knowledge should be forbidden because it is inherently harmful, both by lowering self-respect and by threatening equality.

Second, suppose that the cumulative results of scientific investigations established fatalism, the view that we are entirely determined in ways that remove free choice. This knowledge would "destroy our capacity to function as moral agents."[6] According to Rescher, here is a situation where knowledge would have such bad consequences (applications) that we should refuse to believe it, regardless of how much evidence favors it. In effect, however, his view is that merely possessing this knowledge would be inherently corrosive of morality. Hence, such knowledge is inherently objectionable on moral grounds.

Third, turning to individual truths rather than general knowledge, sometimes we cannot handle horrific truths, truths that destabilize us and undermine our ability to function.[7] Depending on the individual, this information, which undermines "a human mode of existence," might include the specific day we will die, the duration of our marriage, exactly what friends and family think of us, when an asteroid will destroy the earth, and any number of facts that puncture the illusions which keep us happy. In short, "some information is simply not safe for us—not because there is something wrong with its possession in the abstract, but because it is the sort of thing we humans are not well suited to cope with."[8]

Rescher attempts to construe these examples as concerned with dissemination of knowledge (access) and abuse of knowledge (application). Yet the examples seem to qualify as instances where the mere possession of knowledge is morally objectionable on moral grounds. When knowledge is so inherently destabilizing that upon learning it we fall apart psychologically and morally, then merely possessing the knowledge might well be undesirable for us as individuals, at least in that regard. It does not follow, however, that there should be a prohibition of science in pursuing knowledge about deadly asteroids, the durability of marriages, or any number of other topics that might yield distressing results for individuals and groups—contrary to what Rescher seems to recommend.

To be sure, Rescher is difficult to interpret. For one thing, especially in discussing the last set of examples (which concern individuals' beliefs rather than scientists' investigations), he seems to suggest that possessing the knowledge is imprudent rather than immoral. He understands morality narrowly, as consisting solely of responsibilities to other people. In contrast, I agree with Kant that morality includes duties to ourselves to maintain our rational capacities and abilities to function. Using this Kantian conception of morality, it is morally problematic to possess knowledge which permanently destroys our ability to cope, and undermines a human mode of existence. In addition, when knowledge undermines our ability to function, it thereby undermines our ability to meet responsibilities to others, and hence becomes a moral matter.

For another thing, when Rescher says that merely possessing knowledge is never undesirable (contrary to what he says about his examples), he seems to have in mind truth rather than knowledge. Truth is a feature of statements about the world, whereas knowledge is a mental state—truth as believed by persons. Rescher's main point seems to be that truth, in the abstract, is never undesirable. The relevant issue, however, is whether possessing knowledge is ever inherently bad. As we have seen, what he says about his examples seems to imply that merely possessing knowledge is sometimes inherently bad, contrary to his thesis. In any case, no clear line separates the mere possession of knowledge from its distribution and application. If we sink into a suicidal depression upon learning from our physician that we only have a month to live, is our knowledge harmful because of its sheer possession or because of how we apply it to our lives?

Let us skirt the issue of distinguishing possession and application of knowledge. The key issue, it is now clear, concerns possessing meaning-subversive knowledge (as distinct from the misuse of knowledge in further pursuits). Meaning subversion occurs when knowledge undermines the values and practices that give life meaning. Rescher's third set of examples remind us that individuals might be better off not knowing certain things. Our interest, however, is more with the impact of creative science on large groups of individuals or on entire societies, as with the impact of evolutionary theory on Victorian society. When, if ever, should science be forbidden because it threatens to uncover meaning-subversive truth?

The scientific ethos says never, and I agree. I very strongly doubt that science will ever establish fatalism, and I suspect that any discoveries about differences in human intelligence will prove interesting rather than threatening. Nevertheless, scientists should not be forbidden from pursuing such issues, and intellectual honesty requires being open to what science discovers. Rescher apparently disagrees, for what he says about fatalism and the intelligence of ethnic groups implies that some knowledge should be forbidden because it is subversive of widely held values that give meaning to lives of many individuals. In celebrating scientific inquiry and creativity, then, Rescher modifies the scientific ethos more dramatically than initially appears. In his view, some knowledge is inherently undesirable to possess, and myriad moral, religious, and other values need to be used to place limits on the creative pursuit of knowledge. Before responding to his view more fully, let us take account of a similar view stated by Philip Kitcher, who places the issue of forbidden knowledge within a democratic setting.

Kitcher: Democratic Constraints on Creativity

In *Science, Truth, and Democracy*, Kitcher is more consistent and direct in substantially modifying the scientific ethos. He quickly dispatches the notion that all truth and knowledge are inherently valuable. Only *significant* truth is valuable, where significance consists in either practical or epistemic usefulness.[9] Practical usefulness includes technology and other applications of knowledge in everyday life. Epistemic (theoretical) usefulness is the value of knowledge in understanding the world, in combining with other knowledge to form unifying perspectives, and in advancing new inquiries. Epistemic value is what we mean in saying that (some) knowledge has intrinsic value, irrespective of its practical applications.

The source of epistemic value is our natural curiosity, which leads us to ask questions and to find aspects of our environment "salient or surprising."[10] Yet, Kitcher emphasizes, there are many other important values besides natural curiosity, including those of morality, politics, and religion. All these values provide social stability, personal meaning, and direction for technological development. In general, epistemic value does not stand apart from practical concerns, much less above them. Kitcher resolutely opposes "any theology of science that would insulate inquiry against moral and political critique."[11]

So far, so good. We can all agree that moral and other values play important roles in shaping the direction and procedures of scientific research. Certainly societies have a moral right to decide how their money is invested. That right, however, can be exercised in desirable and undesirable ways. It becomes clear that Kitcher thinks that the possession and pursuit of knowledge might justifiably be forbidden by dominant moral, political, and religious values within a society. Contemporary scientific investigation is expensive, and it is funded primarily by government and corporations. Democracies provide concrete procedures for ensuring that the public's money is spent on matters of importance to them, without waste. Importance is assessed partly in practical terms in light of dominant values and partly in terms of democratic values themselves. Those values can reasonably forbid the pursuit and hence the possession of particularly objectionable knowledge.

Some knowledge might threaten meaning-giving values widely affirmed within particular democracies, including democratic values themselves. Imagine the social destabilization that would occur if any of the following propositions turned out to be true (the first two of which echo Rescher's cases). (1) Free choice is an illusion, and all choices are fully determined by physiological factors beyond our control. (2) Genetic differences limit the abilities of members of minority groups who in the past were socially discriminated against, and there are no ways to compensate for these differences. (3) Love and altruism are illusory and based entirely on unconscious egotistical manipulative strategies. (4) Stable human relationships depend on shared illusions about our place in nature, including myths about religious faith.[12]

Kitcher devotes special attention to the second example—genetic differences pertaining to ethnicity.[13] Detailed studies of IQs and genetically fixed potentials of disadvantaged and discriminated-against groups might bolster traditional stereotypes that harm minorities. The inquiries often provide ambiguous evidence that can be used for racist or sexist purposes. In this way, they undermine the democratic ideal of valuing all people equally. They also undermine the self-respect and self-worth of members of disadvantaged groups, and hence their ability to pursue meaningful lives.

Democratic control over science, Kitcher adds, does not imply a crass tyranny in which a majority vote shapes the fate of particular lines of inquiry.[14] Ideally, informed participants educate each other and respond respectfully to competing preferences, reaching "tutored preferences" and rational compromises. In practice, however, we must live with flawed institutions and fallible procedures in allocating funds, both at the macro level of national funding and the micro level of decision making within corporations and universities. In any case, scientists should accept the responsibility to educate other citizens about the promise and perils of their research.

Taken together, Kitcher's and Rescher's arguments are compelling but not convincing. Both of them go too far in allowing conventional moral, religious, and political values to constrain scientific creativity. Ideally, moral, religious, and political values should be revised in light of free scientific inquiry.

Reaffirming the Scientific Ethos

Science has always been politicized, and today more than ever it depends on funding that comes with strings attached. (If knowledge is power, so is money.) In recent years, however, the degree of politicization has become alarming. Occasionally political bias comes from liberals, for example in opposing genetic studies of I.Q. that they worry could threaten egalitarian ideals. More often it comes from conservatives, on two fronts.[15] One front is the culture wars. Thus, conservatives have vigorously opposed embryo and stem cell research, and they have attempted to subvert the teaching of evolution by endorsing Intelligent Design as a viable alternative explanation. The other front is the slanting of science for business purposes, most notably in downplaying global warming, but also in supporting selective industries such as big tobacco.

Kitcher recognizes that the history of science demonstrates "the possible value of subversive truth" in overthrowing oppressive value perspectives.[16] During transitions caused by scientific advances, some individuals and groups might suffer, but in the long run societies usually prosper. Nevertheless, he thinks there can be justification for blocking scientific inquiry when it threatens to subvert meaning, as in his four examples. If what he intends are temporary moratoriums, until safeguards are put in place, I agree. But like Rescher, he seems to intend much more, and that is where I disagree.

Restricting science on the grounds that it threatens to be meaning subversive is itself meaning subversive. It erodes some of the deepest and most durable of all sources of meaning—namely, truthfulness and (significant) truth. These values provide part of the rationale for democratic procedures in the first place, procedures aimed (in part) at fostering the free pursuit of knowledge. Although not all truth is inherently valuable (some is trivial), truthfulness is inherently valuable. So are intellectual honesty, intellectual courage, curiosity, and moral creativity. We should seek to make our moral, religious, and political outlooks compatible with these values. Rather than beginning with the primacy of conventional moral and religious sources of meaning, we should begin with intellectual virtues—which are moral virtues.

Of course, other virtues matter too. Love, friendship, compassion, respect for persons, and justice are as important as intellectual virtues. I also agree that priorities in funding research should be established through rational and democratic procedures of the sort Kitcher outlines. But those procedures ought to keep paramount the intellectual virtues, resisting the politicizing of science in every way possible. Democracies have a right to undermine the ideal of truthfulness, but it is usually wrong for them to do so. Any prohibitions on the free pursuit of science should be rare and temporary, and based on clear and present dangers in acquisition, application, and access. Threats of meaning subversion are not justifications for blocking creative inquiry. None of Rescher's and Kitcher's examples of suppressing creative inquiry on the grounds of meaning subversion strike me as convincing. The goal should be to seek truth using the best scientific methods, and then adjust our moral, political, and religious views accordingly.

Forbidding potentially meaning-subversive inquiries can only generate a backlash. For example, banning inquiries into genetic differences bearing on social inequalities tends to encourage stereotypes by suggesting the official ban serves to cover up unpleasant truths. For this reason, Kitcher and Rescher agree that research should rarely be made illegal, except on grounds of acquisition, access, and application. They appreciate that constraints on scientific inquiries take many forms beyond legal prohibitions: for example, refusal to fund research; peer pressure and regulation within scientific groups; social pressures from outside the scientific community; conscientious choices by individual scientists. Outright legal prohibitions should be limited to situations where extremely dangerous misuses of knowledge are likely. This leaves Kitcher with what he admits is a somewhat "gloomy" conclusion that some lines of scientific inquiry are immoral, ought not to be funded, and should be strongly discouraged, but they should rarely if ever be officially banned.[17]

There are additional pragmatic considerations which bolster the case against allowing parochial pressures to impede science. In a competitive global economy, nations that curtail research will be left behind.[18] Moreover, we might never gain the comprehensive understanding we need to solve many practical problems, such as crime and inequality in education, if we allow politics to block creative inquiry. These pragmatic considerations involve moral reasons, and we need a conception of science as the ally of moral responsibility, not its adversary.

Kitcher fails to provide such a conception of moral responsibility, based on objectively defensible values. Although he provides an insightful defense of the objectivity of truth, he dismisses the possibility of objective moral values.[19] Yet, if values cannot be defended according to their reasonableness, Kitcher's own reliance on democratic values is undermined. Granted, there is no canonical hierarchy of values that all rational persons should affirm. Nevertheless, there are multiple values which are objectively warranted, and which are open to alternative interpretations by reasonable persons. This is moral pluralism, not subjectivism of the sort Kitcher embraces.

In sum, the scientific ethos has served us well, both through technology and in understanding the world. It needs to be balanced with other vital ideals, including justice and compassion. Nevertheless, we should not forbid the pursuit of knowledge simply because it threatens conventional sources of meaning. There is no likelihood that everyone will endorse a strong scientific ethos, but that does not mean we should regard it as merely one among many outlooks competing for a hearing within democracies. The commitment to truth reaches deeper, and enters into justifying and sustaining democracies. Sustaining this faith in the scientific ethos requires creative leadership and teaching, to which I turn in the following chapters.

Notes

1. Deborah G. Johnson, "Forbidden Knowledge and Science as Professional Activity, *Monist* 79, no. 2 (1996): 197–217. See also Roger Shattuck, *Forbidden Knowledge* (San Diego: Harcourt Brace & Company, 1996).
2. Carl Sagan, *The Demon-Haunted World: Science as a Candle in the Dark* (New York: Ballantine Books, 1996).
3. Nicholas Rescher, *Forbidden Knowledge and Other Essays on the Philosophy of Cognition* (Dordrecht: D. Reidel Publishing, 1987), 9.
4. Rescher, *Forbidden Knowledge*, 15, 16.
5. Rescher, *Forbidden Knowledge*, 5.
6. Rescher, *Forbidden Knowledge*, 10.
7. Rescher, *Forbidden Knowledge*, 5.
8. Rescher, *Forbidden Knowledge*, 9.
9. Philip Kitcher, *Science, Truth, and Democracy* (New York: Oxford University Press, 2001), 65.
10. Kitcher, *Science, Truth, and Democracy*, 81.
11. Kitcher, *Science, Truth, and Democracy*, 182, 199–200.
12. Kitcher, *Science, Truth, and Democracy*, 151–52.
13. Kitcher, *Science, Truth, and Democracy*, 96–100.
14. Kitcher, *Science, Truth, and Democracy*, 117–135.
15. Chris Mooney, *The Republican War on Science* (New York: Basic Books, 2005).
16. Mooney, *The Republican War on Science*, 149.
17. Kitcher, *Science, Truth, and Democracy*, 105–107.
18. Cf. Peter Singer, "Ethics and the Limits of Scientific Freedom," *Monist* 79, no. 2 (1996): 227–28.
19. Kitcher, *Science, Truth, and Democracy*, 162–65.

CHAPTER EIGHT

~

Leadership

Morally creative leadership means leading in new and morally valuable ways. It can also mean, more specifically, leading for moral creativity—that is, fostering morally desirable forms of creativity within a group. The two meanings overlap, and both are relevant to leadership in science. After clarifying what leadership is, I illustrate morally creative leadership in several contexts: government projects, university labs, and corporations. In doing so I emphasize that what counts as morally creative leadership turns on our value judgments. Throughout, and especially in concluding, I comment on the interplay of character and institutional practices in creative leadership.

Defining Leadership

Broadly conceived, leadership is shown in all creative activities that inspire or guide others. In this broad sense, Darwin and Einstein manifest indirect leadership within the wider culture, as well as within science.[1] Here, however, we are concerned with direct leadership, that is, exerting influence through personal interaction, for example, by holding positions of organizational authority, persuading a group to adopt a plan of action, or fostering cooperation in working toward shared goals.

Definitions of leadership might be normative (value laden, prescriptive) or descriptive (value neutral). Normatively, a leader is someone who markedly influences a group in a desirable manner, effectively and ethically guiding it toward valuable goals. In this sense, leadership is a virtue, and to call someone a "true leader" is to praise him or her for the positive influence he or she has on others. Normative definitions proliferate whenever the aim is to prescribe and inspire, but they are also familiar in the scholarly literature.[2] For example, James MacGregor Burns set forth an influential definition of transforming (or transformational) leadership as occurring when "leaders and followers raise one another to higher levels of motivation and morality."[3] Transforming leadership is thus moral, creative, and noncoercive. In this sense, Hitler was not a leader but instead a "terrible mis-leader: personally cruel and vindictive, politically duplicitous and treacherous, ideologically vicious and annihilative in his aims."[4]

In contrast, descriptive definitions do not build in particular values that should guide leaders. According to one descriptive definition, leaders hold positions of leadership, official organizational roles that offer special opportunities for exerting influence. Alternatively, leaders are persons who bring about change within groups, whether the changes are good or bad. These definitions differ because individuals holding positions of leadership might be ineffective and fail to exert influence; conversely, many influential persons do not hold positions of authority over others.

Although my interest is in moral leadership, I find it clearer to begin with a descriptive definition and then invoke moral values independently. To that end, I adopt a version of the last definition: Leadership is a relationship in which some members of a group (leaders) exert a marked influence on other members (followers), usually by articulating or exemplifying purposes that establish a direction for the group. This definition leaves open whether leadership is good or bad in a particular instance, and whether authority and coercion are appropriate in particular situations. It also allows that leaders might not hold positions of authority. Thus, leadership can come from teachers rather than principals, citizens rather than politicians, soldiers rather than officers, and workers rather than managers.

If leaders influence, then morally creative leaders influence in morally desirable ways. Morally creative leadership involves several tiers of value: competence and effectiveness, ethical procedures, and morally valuable results. In addition, luck plays a role, including luck in finding enough people willing to follow one's lead. President Clinton's reduction of the federal deficit is an example of morally creative leadership; his failed attempt to reform American health care is an example of weak leadership. Of course, much leadership is morally mixed or ambiguous, desirable in some ways but not others.

Contexts of Moral Leadership

What we count as morally creative leadership turns on our moral evaluations of means and ends, procedures and outcomes. The point is well illustrated by

Robert Oppenheimer in directing the Manhattan Project that created the atom bomb. Oppenheimer possessed multiple talents. As a scientist, he was singularly well-qualified to direct a technological project of unprecedented complexity. As a manager, he quickly learned to recruit talented scientists and engineers, organize work assignments, resolve countless conflicts, and keep the project on schedule. He was a charismatic figure whose "main influence came from his continuous and intense presence, which produced a sense of direct participation" in everyone.[5] In addition, he brought out the best in each scientist, engineer, and staff member, more through his inspiring intelligence and commitment than by issuing orders.[6] Workers and their families appreciated his empathy throughout the demanding schedule set for the three-year project.

Oppenheimer was effective, influential, but was he a morally creative leader—someone who leads in new and morally valuable ways and who elicits moral creativity in others? In the eyes of some observers, Oppenheimer was morally creative on a heroic scale because the Manhattan Project saved the lives of countless American and Allied soldiers and later, through nuclear deterrence, reduced the likelihood of world wars. For other observers, Oppenheimer was not morally creative because he made possible the horrors at Hiroshima and Nagasaki, not to mention the costly nuclear arms race. For still others, the Manhattan Project is morally ambiguous in ways that carry over to assessing his leadership. Apparently that was Oppenheimer's attitude. Shortly after his initial exhilaration upon completing the Project, he began a "brooding ambivalence" that never altogether left him.[7]

Good character interacts in complex ways with specific circumstances. Biographers regard Oppenheimer's effectiveness at Los Alamos as traceable to his high intelligence, self-discipline, and empathy, but also to his responses to a unique situation. Other aspects of his complex personality could have undermined his leadership if he not held them in check. In different contexts, he had a tendency to be arrogant, patronizing, and caustic toward scientists less stellar than himself. Following the Manhattan Project, during his nineteen years as Director of the Institute for Advanced Studies, his "abrasive streak" and "fierce arrogance" would sometimes erupt with "a ferociousness that startled even his closest friends."[8] Sometimes he had difficulty controlling his impatience. Colleagues speculated that this impatience partly explains why he never conducted scientific research of Nobel-Prize caliber, even though he was otherwise capable of doing so.

Oppenheimer was a charismatic leader—a high-profile visionary with singular gifts to inspire. Charisma might be less helpful in other contexts where leadership is shared and multidirectional, radiating not only from top down but also laterally and from bottom to top, in dynamic and shifting patterns.

To illustrate, consider the director of a space science lab at a large Midwestern university.

The director is a distinguished eighty-one-year-old scientist who has a hands-off management style, the exact opposite of Oppenheimer's approach during the Manhattan Project. He meets informally once a week with the scientists he supervises, usually for lunch, but otherwise he gives scientists maximum freedom to pursue their research, within the mission statement of the lab and funding restrictions. He creates an environment where researchers accept responsibility and feel trusted, supported, and stimulated to be creative. By exercising direction over their work, the researchers essentially become coleaders within the lab. Even the students involved felt "an uncommon sense of ownership over the outcomes of the group's work, and so they became all the more interested, engaged, and motivated."[9] The lab is exceptionally productive and has contributed to designing and operating thirty-five space missions.

The highly structured Manhattan Project and the loosely configured research lab represent two extremes. Leadership in most for-profit corporations falls somewhere between. In *Built to Last*, Jim Collins and Jerry Porras studied highly successful companies that had survived for a century, comparing each with a similar company that was good but not great. They found that the most successful companies remain committed throughout their history to a stable set of core values and purposes, while at the same time striving to innovate. In each instance, profits are very important, but they are kept subordinate to the companies' core values and purposes. Collins and Porras also discovered that charismatic leaders are far less important than is often assumed. Instead, the companies themselves are visionary by maintaining their core values and commitments. Furthermore, when these companies do have charismatic leaders they usually cultivate them from within the organization, again illustrating the importance of institutional practices.

For example, 3M is a company with the wide-scoped mission to solve problems innovatively—creativity is built into its goals and values. Other core values include respect for employees' personal growth and initiative, and product quality and reliability.[10] For a century, 3M maintained a climate of innovation through shared initiative and shared leadership, rather than relying on a charismatic CEO. Managers are expected to maintain a climate in which valuable new ideas for products are given a chance to develop.

Thus, a young engineer with a promising idea is sponsored by an "executive champion" who, functioning more as a mentor and coach than a boss, recruits a "new venture team."[11] The team stays together throughout the development of a new product. If the project is successful, engineers advance in pay and title, from project engineers to product line engineering managers to research and development managers of a department. If the project fails, engineers return to their previous level of compensation, thereby providing a safety net of job security that encourages risk taking. Successes are celebrated, and 3M generates over a hundred new products each year. Post-it Note Pads are just one of those products. The idea, which grew into a $100 million product line, came from a 3M employee who sang in a church choir and was annoyed by how slips of paper used to mark pages in the hymnals were always falling out. Another invention was masking tape, the brainchild of a 3M lab technician who heard from sales personnel that auto painters were having difficulty keeping the colors of two-toned cars from running together.

Taken together, the Manhattan Project, the university space lab, and 3M Corporation present a rather idyllic picture of support for creative people. What about the harsher side of capitalism that Joseph Schumpeter called "creative destruction"? Can't negative incentives support innovation? Of course they can, at least for a while, although here we are likely to disagree in our assessments of morally desirable means.

Consider General Electric (GE), which was the first company during the 1980s to voluntarily restructure in ways that came to be called "reengineering."[12] Jack Welch, who held a Ph.D. in chemical engineering and worked at GE for twenty years, was appointed CEO at a time when the company was doing well as the ninth most profitable company among the Fortune 500. The country was in a recession, however, and Welch was convinced that GE needed to be more competitive by becoming faster and more agile. Welch was aggressive in reducing its work force and dismantling excessive bureaucracy. Within five years he laid off 81,000 people and sold businesses with another 37,000 people. One in four GE employees lost their jobs. Steps were taken to help fired employees find new jobs, but the trauma to them and to their families was enormous. In addition, steps were taken to keep the most qualified people, offering increased pay and stock incentives to the top 20 percent, cultivating the core 70 percent, and removing the bottom 10 percent—a system that every executive was required to use in making annual evaluations. Managers themselves were evaluated and advanced, or fired, according to GE's "four Es" of leadership: "very high *energy* levels, the ability to *energize* others around common goals, the *edge* to make tough yes-and-no decisions, and finally, the ability to consistently *execute* and deliver on their promises."[13]

To his enemies, Welch was "neutron Jack," who destroyed people while leaving buildings and resources standing. To his supporters, he was a miracle worker. He transformed GE's bureaucracy into a meritocracy where creativity was rewarded and poor performance penalized. Stockholders were delighted, and Welch became an international spokesperson for creative change. Welch is convinced that he maintained GE's core values of customer benefit, product excellence, and valuing employees. In his eyes, to value employees is to treat them with dignity, place them in the right jobs where they can succeed, listen to their ideas, evaluate them with honesty and integrity, and reward their creativity.[14] At the same time, he insists, these values must be pursued within a corporation whose primary social responsibility is to create wealth in order to stay strong and competitive, provide a reasonable return to investors, and pay taxes to help the country.

Welch represents a world where creativity in science and technology are thoroughly subordinate to economic values. That subordination is increasingly the norm in science-based service professions such as health care, and even in academia.[15] This economic-dominated world is more productive but also alarming—creative destruction, indeed. The future of leadership will require better ways to integrate economic concerns with scientific creativity.

As a last illustration, consider leadership in building an ethical climate within technological corporations. Since its inception, Texas Instruments (TI) was proud to set a high moral tone for itself and the industry, but during the mid-1980s it sought a more formal approach in light of public perceptions of corruptions in the defense industry.[16] It established an Ethics Office and appointed one of its vice presidents, Carl Skooglund, as the first ethics director. Skooglund created a communications channel for conveying ethical concerns from employees all over the world, while preserving confidentiality. He developed an extensive ethics education program and wrote a series of *Cornerstone* publications discussing specific ethical issues and illustrations.

Then, around 2000, TI underwent major changes which included a new CEO and selling twenty subsidiaries over a two-year period.[17] Suddenly TI had more employees outside the United States than inside, and difficulties arose in implementing the detailed *Cornerstone* guidelines. In response, Skooglund dramatically simplified the ethics program around three core values, each with two components: Integrity (respect persons, be honest); Innovation (learn and create, be bold in taking reasonable risks); and Commitment (accept responsibility, commit to TI). It is noteworthy that under "innovation" TI endorses creativity as a moral value. Skooglund and TI showed moral creativity in establishing and then modifying an innovated ethics program.

Individual Character and Institutional Practices

Ironically and importantly, leadership cannot be understood by focusing exclusively on leaders. This narrow focus gives rise to the Great Leader Theory—the idea that the world turns primarily on the presence or absence of great leaders. Often this theory is assumed when we, as members of organizations or the public, oscillate between idealizing and demonizing leaders. Most scholars have abandoned the Great Leader Theory because it neglects the role of broader institutional pressures, economic and social forces, and luck. As the above examples suggest, morally creative leadership comes from good character interacting with good institutional practices. Let us comment further on each.

Character, to begin with, is not all or nothing, and myriad virtues can be relevant to good leadership, depending on the situation. Leadership studies take account of this complexity by using language that often conceals values beneath scientific sounding terms of psychology and the social sciences. Daniel Goleman's psychological theory of emotional intelligence is refreshing because it makes the virtues somewhat more salient, although still not as completely as they should be.[18]

In the original version of his theory, Goleman identifies five dimensions of emotional intelligence: self-awareness, self-regulation, motivation, empathy, and social skills. He then specifies five competencies under each dimension, for a total of twenty-five emotional competencies. For example, self-regulation includes self-control, trustworthiness, conscientiousness, adaptability, and innovation. Again, empathy includes understanding others, helping others develop, a service orientation, cultivating diversity, and political awareness. Most of these competencies either are or allude to virtues. In asides, Goleman admits as much: "There is an old-fashioned word for the body of skills that emotional intelligence represents, *character*."[19]

We can learn much about good character by examining bad character.[20] In a wide-ranging study of leaders, Barbara Kellerman identifies several sources of *un*creative leadership: incompetence (lacking requisite knowledge or skills), rigidity, intemperance (lack of self-control), callousness (disregards the needs of the group), corruption (lies, steals, cheats), insularity (disregards the needs and rights of those outside the organization), and evil (commits atrocities and other severe harm).[21] Some of these terms name character defects, for example, intemperance and callousness. Others allude to vices, either directly (e.g., corruption) or indirectly (e.g., insularity). In elaborating on these types of ineffective leadership, Kellerman readily invokes terms denoting more specific vices, such as greed, disloyalty, cowardice, and selfishness.

Kellerman is aware that invoking character traits has long been out of fashion in psychology, which aspires to be a value-neutral science, although the "positive psychology" movement is changing that with its emphasis on desirable features of personality and character. She emphasizes that leadership is a moral, not a value-neutral, topic. She also calls attention to "the situation, the nature of the task at hand, and of course the followers."[22] Character and context should be studied in tandem. Furthermore, Kellerman is sensitive to "the paradoxes of leadership—leaders who are, for example, corrupt and effective at the same time."[23] Good leaders are both effective and ethical, and both features have multiple dimensions. Usually leaders are effective and ethical in some ways but not all.

Turning to institutional practices, a 1996–1999 study of scientists, physicians, and technicians identifies three main practices that discourage creativity, beyond simple dishonesty and lack of organization: abusive and wholly negative evaluation of subordinates, especially in publicly humiliating them; unwillingness to confront and deal with conflict, thereby allowing problems to fester; and being "selfish, exploitive, dictatorial, or disrespectful."[24] Conversely, effective leaders are "caring, compassionate, supportive, enthusiastic, motivating"; "a good role model, mentor, and coach"; and are competent in both technical matters and managerial abilities to organize, listen, communicate effectively, and resolve conflicts.[25] These qualities, which combine virtues and valued skills, lead to motivating and eliciting the best work from research teams.

Earlier, G. A. Steiner identifies some ways to encourage creativity within organizations.

> Open channels of communications are maintained. Contacts with outside sources are encouraged. Nonspecialists [in addition to specialists] are assigned to problems. Ideas are evaluated on their merits rather than on the status of their originator. Management encourages experiments with new ideas rather than making 'rational' prejudgments. Decentralization is practiced. Much autonomy is allowed professional employees. Management is tolerant of risk-taking. The organization is not run tightly or rigidly. Participative decision making is encouraged. Employees have fun.[26]

Steiner also identifies ways to kill creativity: Discourage risk taking, new ideas, open communication inside and outside the corporation, nonconformity, and enjoyment, and fail to provide recognition and resources.[27]

In understanding institutional practices, we should distinguish between individual and group creativity.[28] Especially in corporate settings, as well as large research projects in universities, creativity is increasingly group creativity. That is, valuable new outcomes emerge from, and can only be ascribed to, an entire group rather than any one individual. Either way, whether in fostering individual or group creativity, attention needs to be paid to institutional contexts and structures that promote creative communication and collaboration.

As just one source of complexity in promoting group creativity, consider Irving Janis's famous study of groupthink. According to Janis, "groupthink involves nondeliberate suppression of critical thoughts as a result of internalization of the group's norms, which is quite different from deliberate suppression on the basis of external threats of social punishment."[29] Group creativity thrives on healthy conflict and controversy that generates alternative perspectives. In contrast, groupthink deadens disagreement and dampers the range of alternatives. Groupthink is insidious because it operates informally, seemingly as integral to moving groups toward a consensus. It can do so even when "brainstorming" is used, given how that technique leads quickly into judgments about which ideas are worth pursuing, judgments often made by either dominant individuals or groups with shared assumptions that should be challenged.

In short, moral creativity in science and technology is not the exclusive domain of isolated individuals. Nevertheless, morally creative leaders make vital contributions, both those who lead in morally creative ways and those who specifically foster creativity in others. The same is true of morally creative teachers, to whom we turn next.

Notes

1. Howard Gardner, *Leading Minds* (New York: Basic Books, 1995), 6.
2. Joseph C. Rost, *Leadership for the Twenty-First Century* (Westport, CT: Praeger, 1991), 140.
3. James MacGregor Burns, *Leadership* (New York: Harper and Row, 1978), 20.
4. James MacGregor Burns, Forward to *Ethics, the Heart of Leadership*, 2d ed., ed. Joanne B. Ciulla (Westport, CT: Praeger, 2004), xii.
5. Victor Weisskipf, quoted by Kai Bird and Martin J. Sherwin, *American Prometheus: The Triumph and Tragedy of J. Robert Oppenheimer* (New York: Alfred A. Knopf, 2005), 277.
6. Bird and Sherwin, *American Prometheus*, 218.
7. Bird and Sherwin, *American Prometheus*, 335.
8. Bird and Sherwin, *American Prometheus*, 386.
9. Charles Hooker and Mihaly Csikszentmihalyi, "Flow, Creativity, and Shared Leadership," in *Shared Leadership: Reframing the Hows and Whys of Leadership*, ed. Craig L. Pearce and Jay A. Conger, (Thousand Oaks, CA: Sage Publications, 2003), 226.

10. Jim Collins and Jerry I. Porras, *Built to Last* (New York: Collins Business, 2002), 68, 225.

11. Thomas J. Peters and Robert H. Waterman, *In Search of Excellence: Lessons from America's Best-Run Companies* (New York: Warner Books, 1982), 224–34; and Thomas J. Peters, "A Skunkworks Tale," in *Managing Professionals in Innovative Organizations*, ed. Ralph Katz (Cambridge, MA: Ballinger Publishing Company, 1988), 433–41.

12. Jack Welch, with John A. Byrne, *Jack: Straight From the Gut* (New York: Warner Books, 2001), 125.

13. Welch and Byrne, *Jack*, 158.

14. Welch and Byrne, *Jack*, 168, 190, 381–99.

15. Derek Bok, *Universities in the Marketplace: The Commercialization of Higher Education* (Princeton, NJ: Princeton University Press, 2003); and Jennifer Washburn, *University Inc.: The Corporate Corruption of Higher Education* (New York: Basic Books, 2005).

16. Francis J. Aguilar, *Managing Corporate Ethics* (New York: Oxford University Press, 1994), 120–35.

17. D. Driscoll and W. M. Hoffman, "From Volumes to Three Words: Texas Instruments," in *Business Ethics*, ed. W. M. Hoffman, R. E. Frederick, and M. S. Schwartz (Boston: McGraw-Hill, 2001), 623–28.

18. Daniel Goleman, *Emotional Intelligence* (New York: Bantam, 1998); and Daniel Goleman, Richard Boyatzis, and Annie McKee, *Primal Leadership: Realizing the Power of Emotional Intelligence* (Cambridge, MA: Harvard Business School Press, 2002).

19. Goleman, *Emotional Intelligence*, 285.

20. See Terry L. Price's *Understanding Ethical Failures and Leadership* (New York: Cambridge University Press, 2006).

21. Barbara Kellerman, *Bad Leadership* (Cambridge, MA: Harvard Business School Press, 2004).

22. Kellerman, *Bad Leadership*, 19.

23. Kellerman, *Bad Leadership*, 47.

24. Alice M. Sapienza, *Managing Scientists: Leadership Strategies in Scientific Research*, 2d ed. (New York: Wiley-Liss, 2004), 6.

25. Sapienza, *Managing Scientists*, 7–8.

26. G. A. Steiner, *The Creative Organization* (Chicago: University of Chicago Press, 1965). Summarized by Albert Shapero, "Managing Creative Professionals," in *Managing Professionals in Innovative Organizations*, ed. Ralph Katz, 216.

27. Shapero, "Managing Creative Professionals," 219–20.

28. Cf. Wendy M. Williams and Lana T. Yang, "Organizational Creativity," in *Handbook of Creativity*, ed. Robert J. Sternberg (New York: Cambridge University Press, 1999), 373–91; and Vera John-Steiner, *Creative Collaboration* (New York: Oxford University Press, 2000).

29. Irving L. Janis, "Groupthink," *Psychology Today* (November 1971), 272.

CHAPTER NINE

~

Teaching

Creative teachers bring about new and valuable outcomes in the form of student learning. In science, learning includes understanding scientific truths and procedures, appreciating the intellectual and social importance of science, and preparing to cope in a technological world. At a more practical level, however, controversies arise concerning what counts as valuable outcomes. Thus, creationism evangelists who teach children to appeal to the Bible in rejecting evolution might be teaching something new, in the sense of beliefs outside the scientific mainstream.[1] Yet, judged by the standards of science, that aspect of their teaching is not valuable and hence not creative.

Creative teaching might consist in using new and valuable approaches, for example, in finding innovative ways to engage diverse populations of students in rural and urban settings. Often the best way to achieve these outcomes is to teach *for* creativity, that is, to foster creativity in students. Creativity can also occur through improving educational institutions and curricular practices that ultimately contribute to learning.

Creative Approaches to Teaching

Science teachers help students master large quantities of information and a variety of skills. In K-12 education, most teachers have limited authority to determine what the information and skills are. Yet, once goals are specified by schools, districts, states, and national standards, benchmarks are established against which creative teaching can be assessed, both as furthering those goals and exceeding them. The need for creative teaching is acute in many rural and urban settings where parents and culture do not take for granted the importance of studying science. Let us consider several examples.

Jaime Escalante was an immigrant from Bolivia who chose teaching mathematics over more lucrative jobs in industry because he loved his subject and cared about students.[2] Given the option of teaching in wealthier school districts, he chose instead to work at Garfield High School in East Los Angeles because he wanted to engage Mexican-American students. At Garfield 95 percent of students were Latino and 80 percent came from families below the poverty line. Expectations were low and drop-out rates high. In response Escalante set unusually high standards in all his classes. He also created an advanced placement calculus class and adopted the goal of having all students pass the national exam with a score of 3 or higher, in order to enhance their chances of attending college. Motivating students with a combination of stimulating classes and personal attention, he pushed them to attend extra classes during lunch hour and after school. He also convinced parents that their children's efforts would increase their chances of graduating from high school and gaining entrance to college. His first Advanced Placement class was so successful that national administrators assumed his students had cheated and required them to retake the exam. When it became clear that his students were learning at a level unprecedented at Garfield, he was placed in charge of Garfield's math program.

A second example is a team of professors and doctoral students at Columbia University who worked with an after-school program for teenagers living in a temporary urban housing shelter.[3] The youths were disadvantaged in virtually every respect, but the teachers emphasized their power to contribute to their community. Together, students and teachers explored a number of possible projects before settling on a plan to develop a community garden on an abandoned lot used for dumping garbage and selling drugs. Naming their group REAL—Restoring Environments and Landscapes—the students planned and developed a vegetable garden, selling its produce to the local community. In addition, they built a stage and drew wall murals to attract community events. Working in teams, the students took measurements, drew alternative designs, and researched relevant ecology, including the basics about how plants grow. In cleaning up the lot students learned about responsible trash collection and recycling. And by interacting with members of the local community, they developed skills in oral and written communication.

A third example is a group of middle and high school teachers in a rural setting who had their classes study the Northwest's Henderson Creek watershed, which drains into the Pacific Ocean.[4] Henderson Creek was polluted from farming and from an expanding suburbia that included an industrial park. In places the creek had been rerouted into straighter lines to make room for housing, but at the cost of losing wetlands and cleaner water flow. Teachers saw an opportunity for students to move beyond the traditional laboratory setting and to link science to community service. With the help of parents as supervisors, students took regular water samples, devised experiments, learned relevant mathematical models, and conducted Internet research on the connection between river and ocean pollution. In addition to developing traditional science fair projects, they reported their findings directly to the local community through meetings and newspapers.

A fourth example involves collaboration between a science teacher and an auto shop teacher at Northampton High School-East in North Carolina, where most students are African Americans from economically disadvantaged families.[5] Eric Ryan was a science teacher serving his second year in the Teach for America program. Harold Miller was the auto shop teacher who for two decades, following the 1973 OPEC oil embargo, had pursued an interest in electric cars as a way to reduce dependency on foreign oil. A county education administrator urged them to work together with a group of students to enter a newly established contest for Mid-Atlantic States. The contest consisted in transforming a gas engine car into an electrically powered vehicle. Student groups would then compete at the Richmond, Virginia, NASCAR track in four categories of driving, including an energy efficiency run, a distance competition, and two categories of design, along with an oral presentation on the science involved. Northampton competed against schools from far more affluent areas, and won. In the distance competition, for example, a student drove sixty-four miles during the allotted ninety minutes, and in the final nine laps she was the sole driver remaining on the track, all the other cars having run out of power. The students' success led Ryan and Miller to develop a team-taught course on electric vehicles. They and their students created a tradition of excellence in future years of competition, and renewed pride and increased resources for the entire school.

These examples remind us that teaching has a highly personal dimension. Teachers are not robots who mechanically transmit information. They are professionals who build on their personal strengths and commitments to challenge students and facilitate learning. There are many ways to teach well and to teach creatively, as J. Myron Atkin and Paul Black note.

> Teachers, like other professionals, are people who stand for certain things and who exemplify attributes of character and personality that are intimately linked to and reflected in their practice. Often they are the teacher's most distinctive

> qualities. . . . One favorite teacher is remembered for being especially passionate about her subject, another for her special way of encouraging students to do their best, another for her generosity in taking extra time with students who are having difficulties.[6]

All creative teachers continually improve their knowledge base and pedagogical skills. They find ways to keep alive their commitments despite inadequate salaries and challenging working conditions. And they remain alert to ways to convey the results of their own intellectual growth into the classroom, including by teaching for creativity.

Teaching for Creativity

What it means to teach for creativity depends on educational level and context. In middle and high school, teaching for creativity typically means fostering new and valuable outcomes relative to individuals (person-relative creativity) or groups (group-relative creativity). Even so, a long-term aim might be to develop skills that enable students to achieve history-relative creativity later on.

In *The Art of Teaching Science*, Jack Hassard illustrates many ways to foster creativity in middle and high schools. For example, he describes a high school chemistry teacher who each day elicits inquisitiveness and creative reflection in her students: "Students are involved in watching minidemonstrations and then trying to figure out what happened, testing the acidity of rain (with cabbage juice as the indicator) . . . and then drawing conclusions based on their own data, conducting microchemistry experiments designed to help them learn chemistry concepts inductively."[7] In addition, the teacher has students participate in international projects, pairing them through the Internet with students in other countries in order to gain international perspectives on environmental chemistry.

Hassard is less helpful, however, in defining what creativity is—that is, what is supposedly being fostered. He defines creativity in terms of special mental processes that differ from those of critical thinking.[8] *Critical thinking*, he tells us, consists in "reasonable, reflective thinking" using such processes as observing, designing experiments, hypothesizing, inferring, measuring, and classifying. In contrast, *creative thinking* is using intuition, imagining, visualizing, and the envisioning of alternatives. The problem with this dichotomy is that the two activities are thoroughly interwoven. For example, formulating hypotheses and designing experiments often require imagining and envisioning alternatives, and vice versa. Hassard seems aware of this interweaving when he says that critical and creative thinking are complementary and together constitute "holistic thinking." Even so, the interweaving is too intimate to allow even an initial separation in terms of mental processes. To be creative in science is to hypothesize, design experiments, make inferences, and engage in other critical thinking *in* creative ways, that is, in ways that bring about new and valuable outcomes. Without constant reference to outcomes, or at least tendencies to produce such outcomes, we cannot understand ideas such as creative thinking, creative approaches, and creative intelligence.

Fostering imagination, flexibility, inquisitiveness, risk taking, and other skills is important, but it needs to be done with an eye to identifying and solving problems and exercising scientific reasoning. As part of fostering creativity, teachers should also convey how science progresses through the cumulative creativity of many investigators, punctuated by the episodic leaps by exceptionally creative individuals. One way to convey this historical sense is by occasionally taking time to recount the personal and social backgrounds of exemplary scientists.

At all levels, teaching for creativity involves conveying values, especially intellectual and other moral virtues and ideals, including intellectual honesty, humility, and courage, as well as love of truth and the ambition to be creative. For the most part, these values must be taught indirectly and conveyed through the enthusiasm or quiet devotion of the teacher. As large amounts of information and explanatory schemas are introduced, they are unfolded with a sense of moral significance. Wherever possible, they are also conveyed with a sense of fun, with pleasure in playing with ideas and conducting experiments.

Having mentioned fun, let us illustrate creative college teaching using a person we have mentioned before. Richard P. Feynman found teaching a forum for creative expression, an opportunity to convey respect for truth and the wonder of nature, and a way to convey the pleasure in solving problems: "I love to teach. I like to think of new ways of looking at things as I explain them, to make them clearer."[9] In his undergraduate courses, his students recall, "the lecture hall was a theater, and the lecturer a performer, responsible for providing drama and fireworks as well as facts and figures."[10] Whether at the undergraduate or graduate level, Feynman taught for creativity. As one student recalls, "He urged each of us to create his or her universe of ideas . . . . It was excitingly different from what most of us had been taught earlier."[11] In teaching creatively, Feynman emphasized, care must be taken to convey the power of science to overthrow superstitions but also its fallibility: "It is necessary to teach both to accept and to reject the past [science] with a kind of balance that takes considerable skill."[12]

Especially in graduate and post-graduate school, teaching for creativity includes challenging students to advance the field of science (history-relative creativity). In the culminating levels, teaching becomes much like coaching.[13] An effective coach transmits information and develops skills in a manner that empowers athletes to excel. At their best, teachers do that as well, and often it becomes their primary role in doctoral and postdoctoral work.

Educational Innovation

Wendy Kopp founded Teach for America, the program in which Eric Ryan participated. As an undergraduate at Princeton University, Kopp became concerned about the inequities in public education. She organized a conference on the topic, during which a question struck her: "Why didn't this country have a national teacher corps of top recent college graduates who would commit two years to teach in urban and rural public schools?"[14] Such a program would appeal to the idealism of graduating seniors who wanted to make a meaningful contribution. In doing so, it would attract innovators to under-resourced schools, where they were most needed. Kopp turned the idea into her senior thesis, and after graduation she found corporate and foundation seed money to put it into practice. Within a year Kopp was the chief administrator and fundraiser for a multimillion dollar program that recruited and placed hundreds of outstanding graduates. During its first thirteen years, Teach for America placed 9,000 teachers who have taught a million children. The majority of these teachers have remained educators and school administrators.

Kopp stands in a long tradition of leaders in education. Other innovators have focused specifically on the math and science curriculum. One enduring challenge is to determine the aims and means in teaching science. Especially in K-12, fundamental aims come from society, whose needs are both changing and contested. Should the emphasis be on transmitting massive amounts of information, instilling appreciation of scientific methods, fostering creativity, nurturing ecological awareness, or responding to the needs of business and national defense? Determining priorities is always subject to controversy, as is determining who has primary authority to make decisions. At one time, writers of textbooks had that authority. Then, especially due to the increased competition with the Soviet Union, leading scientists exercised primary control through the National Science Foundation. Today, state and local governments compete for control, and teachers often have limited input. It is clear that setting the aims and content of science curricula needs to be a shared creative activity.

Another area requiring shared creativity is integrating and coordinating the teaching of various sciences. How much physics, chemistry, biology, geology, computer science, and technology studies are necessary? How will all these subjects be integrated to provide a coherent educational experience? Is it better to blend disciplines or to keep them distinct but coordinated at key points, such as in student projects studying different aspects of energy use or environmental protection? Integration is needed at all levels: during a given school year, between successive school years, and between school levels, especially between high school and entrance into college. It is desirable, for example, to have state universities set minimum entrance requirements, but high school teachers also need to have a say in what is feasible.

Integrating science theory with technology studies is especially important.[15] Science and computer literacy continue to be major concerns, and broader technological literacy is increasingly important. Technology is a natural way to establish the relevance of science, as in the electric car example. Technology is based upon applying science, but it also involves applying distinctive engineering concepts, especially the concept of design, in ways that connect with social science, communication studies, and even philosophy in solving practical problems and serving community needs.

Always there is a challenge in how to balance excellence with equal opportunity.[16] Each reform brings unintended consequences.[17] For example, the push for standardized national tests seems an obvious way to raise the level of competence in science and math. Invariably, however, it provides an incentive to "teach to the test" in order to have as many students as possible pass. The upshot is that teachers' work becomes standardized in ways that can discourage creative initiatives. Creative group projects might be downplayed because they are not easily tested using multiple choice and short answer questions. Funding for field trips might be cut as limited resources are redirected toward testing in standardized areas. And there is an incentive to allow less talented students to drop out rather than to report them as failing national exams.

The deepest problems in K-12 education, the ones calling for the most shared creative responses, reside in recruiting and retaining excellent teachers. Teachers need to be paid much more.[18] Creative political leadership, especially at the state and federal level, are needed to make that happen. Each U.S. president claims to be a leader in education, but so far no president has taken that goal as seriously as military budgets.

University teachers have far more leeway to be creative in how they teach and in teaching for creativity. Even when their courses are integral to broader programs over which they have limited control, they have fresh opportunities each semester to innovate. As just one illustration, consider Jeanette Norden, a cell biologist who teaches neuroanatomy in Vanderbilt University's Medical School. To a significant degree, Norden must "teach to the test" in order to prepare her students for the U.S. Medical Licensing Exam and the test for the National Board of Medical Examiners. During the 1990s, however, she became aware that her students were ill prepared to deal with the human side of medicine. In particular, they had great difficulty helping patients and their families come to grips with death. The curriculum was tight, but she found ways to integrate moral topics about death into her courses. For example, by getting students to read more outside of class, she freed up class time for personal discussions of death and for teaching practical skills in communicating sympathy. Creatively integrating technical and moral matters, she helped her students "confront their own mortality and the frailties of the human condition, a reality in which people do die, and a profession that must care both for healing and for helping people and their families face the inevitable with dignity and peace."[19]

In sum, creativity is as important in teaching science as in doing scientific research. Creative teaching takes many forms, only a few of which I have illustrated. There is also, for example, teaching by writing popular-science books, or creating innovative mass-media programs such as Carl Sagan's *Cosmos* and Jacob Bronowski's *Ascent of Man*. Whereas creativity in research is highly prized, in education many forces work against it, especially inadequate resources and low pay that push talented individuals into other careers.

Notes

1. See Stephanie Simon in "Their Own Version of a Big Bang," *Los Angeles Times* (February 1, 2006), 1, 16.

2. Jay Mathews, *Escalante* (New York: Henry Holt, 1988). The film *Stand and Deliver* portrays Jaime Escalante, who subsequently extended his creative teaching career with a program on public television.

3. Angela Calabrese Barton, *Teaching Science for Social Change* (New York: Teachers College Press, Columbia University, 2003), 138–57. See also *Improving Urban Science Education*, ed. Kenneth Tobin, Rowhea Elmesky, and Gale Seiler (Lanham, MD: Rowman & Littlefield, 2005).

4. Wolff-Michael Roth and Angela Calabrese Barton, *Rethinking Scientific Literacy* (New York: Routledge Falmer, 2004), 22–48, 159–79.

5. Caroline Kettlewell, *Electric Dreams* (New York: Carroll and Graf Publishers, 2004).

6. J. Myron Atkin and Paul Black, *Inside Science Education Reform: A History of Curricular and Policy Change* (New York: Teachers College, Columbia University, 2003), 157.

7. Jack Hassard, *The Art of Teaching Science* (New York: Oxford University Press, 2005), 210.

8. Hassard, *The Art of Teaching Science*, 332.

9. Richard P. Feynman, *The Pleasure of Finding Things Out*, (Cambridge, MA: Perseus Publishing, 1999), 203.

10. David Goodstein, "Richard P. Feynman, Teacher," in *"Most of the Good Stuff": Memories of Richard Feynman*, eds. Laurie M. Brown and John S. Rigden, (New York: Springer Verlag, 1993), 118. Cited by Kay Redfield Jamison, *Exuberance: The Passion for Life* (New York: Alfred A. Knopf, 2004), 238.

11. Laurie M. Brown, "To Have Been a Student of Richard Feynman," in *"Most of the Good Stuff,"* ed. Brown and Rigden, 54.

12. Richard P. Feynman, *The Pleasure of Finding Things Out*, 188.

13. John Janovy, Jr., *On Becoming a Biologist*, 2d ed. (Lincoln: University of Nebraska Press, 2004), 75.

14. Wendy Kopp, *One Day, All Children* (New York: PublicAffairs, 2003), 6, italics removed.

15. For helpful resources, see *Technically Speaking: Why All Americans Need to Know More about Technology*, ed. Greg Pearson and A. Thomas Young (Washington, DC: National Academy Press, 2002).

16. John W. Gardner, *Excellence: Can We Be Equal and Excellent Too?* (New York: Harper & Row, 1961).

17. J. Myron Atkin and Paul Black, *Inside Science Education Reform*, 150.

18. Daniel Moulthrop, Ninive Clements Calegari, and Dave Eggers, *Teachers Have It Easy: The Big Sacrifices and Small Salaries of America's Teachers* (New York: New Press, 2005).

19. Ken Bain, *What the Best College Teachers Do* (Cambridge, MA: Harvard University Press, 2004), 90.

CHAPTER TEN

Good Lives

Properly directed, creativity enriches the meaning professionals find in their work and contributes to the well-being of humanity; misdirected, it leads to scientific misconduct and unsafe technologies. These themes connect to broader issues about good lives. How does the ideal of being creative enter into such lives? What impact does it have on personal life and the integration of personal and professional life? Is creativity connected to wisdom? All things considered, is creativity perhaps overrated?

Dangers to Self and Family

Daniel J. Levinson interviewed ten biologists over several years as part of a study of adult development. Some of his findings are illustrated by a professor at Columbia University to whom he gives the pseudonym John Barnes.[1] At age twelve, Barnes decided to become a biochemist in the hope of making significant contributions to science and humanity. After earning a Ph.D. from Yale, he took a position at Columbia where he earned early tenure. He worked single-mindedly, protecting his long hours in the lab by giving minimal attention to his family and to world events. During one exhilarating week when he was thirty-two, he made more fundamental discoveries than he would in the remainder of his career. Shortly thereafter he was drawn into academic administration, where he also excelled, but eventually he returned to research where he was again creative.

Barnes was nationally and internationally renowned, but not to the extent he desired. Since childhood he dreamed of making momentous discoveries that would shake the foundations of biology and earn him a Nobel Prize. In his early forties, he became disillusioned about his accomplishments. His idealism turned to despair and became mixed with misanthropy: "He wanted to be the hero who gains immortality by saving the world; and at the same time he wanted to destroy the world or run away and let it destroy itself. Along with his strong humanitarian concerns, he had doubts that society is worth saving."[2]

Barnes's idealism was corrupted by egoism and elitism: "I simply can't get away from the thing that really does trouble me, which is an intense desire for kudos. . . . Sure, it's getting something accomplished [that counts], but why does it count? It counts because you get kudos for it from your peers, scientifically."[3] It took several years before his preoccupation with fame began to lessen. Gradually he found renewed satisfaction in his work, even though it did not match his youthful ambitions. He developed a more balanced life in which family, friends, and civic involvements became primary sources of meaning, along with his creative professional endeavors. He also came to appreciate his creative contributions in teaching, mentoring junior colleagues, and leading through administrative positions.

Barnes reminds us that ideals of scientific creativity are rarely embraced with pure regard for truth and understanding. Commitment to truth can be genuine and yet mixed with other aims and motives, including strong desires for recognition. Mixed motives can be desirable by strengthening commitments (chapter 4), although they can also lead to professional misconduct (chapter 6). And even when individuals maintain professional integrity, an excessive ambition to be creative can have harmful side effects on self and family, as Barnes illustrates.

The early years of Barnes's career took a huge toll on his family. He was stunned when his wife of ten years left him for another man. She had often complained to Barnes about his lack of emotional involvement, but he assumed there was no deep problem. Such family matters are beyond the purview of professional ethics as usually understood, but certainly they bear directly on good lives of professionals.

Howard Gardner takes seriously the interaction of personal and professional life in his study of exceptional creators such as Einstein and Picasso. Gardner was struck by how they, "in order to maintain their gifts, went through behaviors or practices of a fundamentally superstitious, irrational, or compulsive nature."[4] In most instances, these behaviors involved sacrificing normal human relationships, depriving themselves of supportive friendships, and abandoning sex with their spouses. Einstein, for example, grew distant from his family, had extramarital affairs, and in general was far less interested in sustaining time-consuming friendships than in pursuing his work. He did, however, care deeply for social justice. He was well aware of these aspects of his personality:

> My passionate interest in social justice and social responsibility has always stood in curious contrast to a marked lack of desire for direct association with men and women. . . . Such isolation is sometimes bitter but I do not regret being cut off from the understanding and sympathy of other men. . . . I am compensated for it in being rendered independent of the customs, opinions and prejudices of others and am not tempted to rest my piece of mind upon such shifting foundations.[5]

What appears to the creative individual as self-sacrifice on behalf of noble pursuits might seem more like arrogant egotism in the eyes of those harmed. As observers, we readily disregard such excesses in creative giants, although we are perhaps less indulgent when other individuals strive to be creative at the cost of harming their families, especially if it involves child neglect and spouse abuse.

The influx of women into all fields of science is a dramatic development since Levinson conducted his studies. The changes taking place are affecting men as well, as greater attention is paid to balancing professional and private life, and sharing family responsibilities. Any generalization about these changes risks reinforcing stereotypes, but sociologist Phyllis Goldberg cautiously offers this observation:

> The rewards typically offered by organizations—higher salary, increased prestige, elevated status, more challenge—do not necessarily reflect what many women deem most necessary and important to them. Conversely, what many women tend to regard as rewarding—collaborative effort, supportive workplace, quality of life, better "fit" between professional and personal life—is rarely included in the traditional package of rewards and benefits.[6]

Goldberg emphasizes that the changes occurring are significant but incremental, as universities and corporations develop more supporting ways to recruit and retain talented women who, as a group, continue to bear primary responsibilities for raising children. The pressing need is to develop supportive structures that take into account the interplay of professions and families in good lives.

Good Lives and Self-Fulfillment

In science as elsewhere, fully good (desirable) lives are morally admirable, meaningful, and happy. First, while all persons have inherent moral worth, not all ways of life are morally desirable. Good lives are morally admirable in terms of general principles of morality such as honesty and justice, and also personal moral commitments and ideals such as pursuing excellence in a profession. As John Kekes summarizes, good lives "combine personal satisfactions, derived from engagement in projects in a manner that reflects one's ideals of personal excellence, and moral acceptability which depends on conformity to the universal, social, and individual requirements of morality."[7]

Second, lives are meaningful when they are enlivened by a sense of meaning rooted in justified values. Subjectively, a sense of meaning is manifested by a rich engagement in life, guided by a set of values, and generating a conviction of being worthwhile. Objectively, meaningful life implies that these values and endeavors are justified and desirable.[8] Although all human lives are valuable, not every life is suffused with a sense of meaning. A newborn infant who dies, a mentally disabled adult with advanced Alzheimer's disease, and a recklessly disordered individual illustrate only a few ways in which a sense of meaning can be absent.

Scientific creativity is linked to a special cluster of ideals understood in terms of scientific discovery and technological design, but these ideals exemplify broader ideals of creativity in contributing to meaningful lives. Especially since the nineteenth century Romantic era, creativity has been regarded as central to individualism. Thus, Matthew Arnold takes it as "undeniable that the exercise of a creative power, that a free creative activity, is the true function of" human beings.[9] Throughout his writings, Nietzsche celebrates creativity as the primary virtue. John Stuart Mill also celebrates "originality in thought and action" in pursuing individuality and happiness.[10] And Shakespeare expressed the key idea most powerfully: "This above all, to thine own self be true."[11]

Authenticity is the current term referring to this combination of creativity and individuality. As Charles Taylor elaborates, authenticity is the idea that "each of us has an original way of being human," and if I fail to be true to myself and to my own originality, "I miss the point of my life, I miss what being human is for *me*."[12] Taylor rightly rejects "the slide to subjectivism" that reduces authenticity to sheer license and self-indulgence. Properly understood, authenticity is anchored in objectively defensible values that provide a "horizon of significance" for our lives.[13] For scientists, these values include the intellectual virtues and especially the ideal of being creative, understood as discovering or inventing new and valuable products.

Insofar as we cherish creativity, we need to identify and nurture our talents.[14] Often we have multiple talents, and there is no general moral requirement to pursue all of them, including those we have no interest in developing. Authenticity is the dominant value guiding choices about which talents to develop in pursuing self-fulfillment. Each of us must discover what our "true self" is, and that discovery can involve person-, group-, or history-relative creativity in the sciences.

Third, fully good lives are happy. Happiness is subjective well-being in the sense of being satisfied overall with our life—living with a sense of well-being, typically based on a positive evaluation of the basic features of our lives as meeting our expectations.[15] This satisfaction need not be passive contentment or continuous pleasure, and usually it is not. For many of us it is found through creative lives that involve the full spectrum and vicissitudes of the emotions, both highs and lows.

From Plato on, philosophers have explored the contribution of happiness to good lives, and today the so-called positive psychologists are giving the topic extensive study.[16] All agree that finding happiness can be elusive, whether in our professional or private lives. According to the paradox of happiness, seeking happiness directly is self-defeating. Instead we should pursue endeavors and relationships that we find inherently worthwhile. Then, with any luck, happiness will come as a by-product.[17] Despite the insight contained in this paradox, it remains true that we choose most of our endeavors and relationships because we believe (or hope) they will contribute to our happiness.

To this end Gertrude Elion offers valuable and unpretentious advice about choosing science as a career: "Choose the field that makes you happiest. There is nothing better than loving your work." She adds that happiness comes along the way, rather than at some culminating end point: "Set a goal for yourself. Even if it is an 'impossible dream' each step toward it gives a feeling of accomplishment. Finally . . . be persistent, don't let yourself be discouraged by others, and believe in yourself."[18] Her advice is noteworthy given that what made her happy was a desire to help alleviate suffering.

What we are best at and what most interests us are not always identical. With this in mind, Alan Gewirth distinguishes two conceptions of self-fulfillment and two corresponding conceptions of happiness.[19] "Aspiration fulfillment" consists in pursuing our strongest desires; accordingly, happiness is the pursuit of what we care about most deeply. In contrast, "capacity fulfillment" consists in unfolding our highest talents and capacities; accordingly, happiness is the development of our most valuable talents and abilities. Gewirth believes both forms of fulfillment and happiness are important, but he favors capacity fulfillment in cases of conflict. Others, like Elion, place primary emphasis on discovering and pursuing what we care about most deeply, assuming it is in the realm of moral permissibility.[20] Successfully integrating our strongest desires and our greatest talents is one of many areas where the pursuit of good lives requires wisdom—practical understanding in the art of living.

Wisdom and Balance

All things considered, is creativity overrated? John Hope Mason, in *The Value of Creativity*, contends that as a society we have gone too far in glorifying creativity. He focuses on artistic creativity, but he implies that his conclusions apply to scientific and other forms of creativity as well. Whether in relentless technological expansion, the excesses of artistic license, or the "creative destruction" caused by global capitalism, we have become "cut loose from the *terra firma* of continuity with the past and have little idea of where we can find a secure mooring."[21] We need greater perspective, a sense of limits, and appreciation of the interdependence and fragility of life. To this end, we should "value the curator as much as (perhaps more than) the creator," so that "our highest aesthetic admiration would not be for heroic individual achievement—a landscape painted by Van Gogh—but rather for a subtle dialogue with nature—a Japanese garden."[22] Creativity is important, but perhaps it is more important to appreciate and conserve what is valuable.

In reply, let us agree that the restraint and enlightened cherishing that Mason calls for are not only compatible with affirming creativity but interwoven with creativity. Creativity is protean and multidirectional. To invert Mason's metaphor, creativity is manifested in both Van Gogh paintings and Japanese gardens, and it is equally needed in reshaping and preserving the environment. Rather than contrasting creators and curators, we should value creativity in curators, who include teachers, science museum directors, and environmental conservationists. We should also think of creativity as conserving and extending valuable traditions of excellence, and direct creativity in science and technology toward supporting good lives and good societies. In short, we should recast Mason's critique of creativity as a call for greater wisdom in how creativity is exercised.

Wisdom is the virtue of exercising sound judgment in living a good life in a balanced, whole, and deep way.[23] It is built on self-knowledge, self-mastery, practical experience, realism about what is possible, and humility about our powers to achieve good. It can also be manifested in imagination and boldness in pursuing opportunities for good. Typically, wisdom is shown in reasonably integrating values and reality, ideals and intransigent realities. Moral values are multiple and often conflict with each other and with other types of values. Wisdom is the "creative intelligence," as John Dewey called it, needed to harmonize competing values to arrive at reasonable responses to challenges.[24]

Wisdom functions as a second-order virtue in examining the values we live by.[25] As such, it enables us to develop our character by strengthening good habits and weakening bad ones. In addition, wisdom is manifested as good judgment in maintaining balance among our myriad endeavors, and in integrating our professional and private lives. John Barnes lacked this kind of wisdom during his early adulthood. He achieved it when he emerged from his mid-life despair by developing perspective on how to balance his professional pursuits with his family and other relationships.

Robert J. Sternberg has extensively studied the relationship between creativity and wisdom. He defines wisdom as balance in living by values that advance the common good.[26] The things needing balance include our interests, long- and short-term goals, competing demands of the circumstances in which we act, and the needs and rights of others. In developing a workable, operational definition of wisdom, Sternberg identifies six features of wisdom: reasoning ability, sagacity, learning from ideas and environment, judgment, expeditious use of information, and perspicacity. Next, he specifies operational criteria that facilitate investigation of each feature. For example, the criteria for sagacity are as follows: "Displays concern for others. Considers advice. Understands people through dealing with a variety of people. Feels he or she can always learn from other people. Knows self best. Is thoughtful. Is fair. Is a good listener. Is not afraid to admit making a mistake, will correct the mistake, learn, and go on. Listens to all sides of an issue."[27]

After developing another list of criteria for creativity, Sternberg studies its relationship to wisdom in a wide sampling of individuals. He has found that creativity and wisdom are not strongly correlated, whether in scientists or people in general. In one respect, wisdom and creativity are negatively correlated: "Whereas the wise person is perceived to be a conserver of worldly experience, the creative person is perceived to be a defier of such experience."[28]

Although creativity does not entail wisdom, Sternberg believes that creativity usually enters into wisdom, especially in solving problems whose solution is not obvious.[29] These solutions require good judgment in discovering valuable new ways to balance and integrate competing moral reasons and values. Accordingly, we might be more likely to find moral wisdom in creative teachers and leaders than in highly specialized problem solvers in cutting-edge research.[30] Echoing Mason's call for appreciating curators as much as creators, Sternberg urges teaching for wisdom along with creativity and intelligence, at all levels of education.

As Sternberg emphasizes, there are multiple criteria for wisdom, some of which might be manifested when others are not. Indeed, Robert Nozick suggests there can be wisdom in not always being fully balanced and prudent: "Completely balanced and proportional judgment might inhibit youth's forceful pursuit of partial enthusiasms and great ambitions, through which they are led to intense experiences and large accomplishments."[31] In science as elsewhere, wisdom involves knowing when to pursue great passions in a focused, even compartmentalized manner, and how to balance those pursuits with other commitments at other times, a comment that echoes Richard Feynman's reflections on his compartmentalized passions during the Manhattan Project.

Perhaps, after all, it is now common sense that wisdom and creativity in science are often disconnected, much too often. As many have observed, the days of thinking creative scientists can solve all our problems are gone. If anything, the present danger is under-appreciating scientists' contributions. At least since the late 1960s, science and technology are seen as Janus-faced forces of destruction and cultural progress.[32] Wisdom about scientific creativity will take account of this moral ambiguity while restoring faith in science and the creative ideal that makes it possible.

Notes

1. Daniel J. Levinson, *The Seasons of a Man's Life* (New York: Ballantine Books, 1978), 67–68, 260–77.
2. Levinson, *The Seasons of a Man's Life*, 275.
3. Levinson, *The Seasons of a Man's Life*, 172.
4. Howard Gardner, *Creating Minds* (New York: Basic Books, 1993), 386.
5. Albert Einstein, as quoted by Gardner, *Creating Minds*, 131.
6. Phyllis Goldberg, "Creeping Toward Inclusivity in Science," in *Women in Science and Engineering: Choices for Success*, ed. Cecily Cannan Selby (New York: New York Academy of Sciences, 1999), 10.
7. John Kekes, *The Art of Life* (Ithaca, NY: Cornell University Press, 2002), 5.
8. Cf. Susan Wolf, "Happiness and Meaning: Two Aspects of the Good Life," in *Self-Interest*, ed. Ellen Frankel Paul, Fred D. Miller, Jr., and Jeffrey Paul (New York: Cambridge University Press, 1997), 209.
9. Matthew Arnold, "The Function of Criticism at the Present Time." Quoted by John Hope Mason, *The Value of Creativity: The Origins and Emergence of a Modern Belief* (Burlington, VT: Ashgate, 2003), 198.

10. John Stuart Mill, *On Liberty* (Indianapolis, IN: Hackett Publishing, 1978), 55, 62. Mill is quoting and endorsing Wilhelm von Humboldt.

11. William Shakespeare, *Hamlet*, Act I, Scene III.

12. Charles Taylor, *The Ethics of Authenticity* (Cambridge, MA: Harvard University Press, 1992), 29. See also Lionel Trilling, *Sincerity and Authenticity* (Cambridge, Ma: Harvard University Press, 1971).

13. Charles Taylor, *The Ethics of Authenticity*, 66.

14. D. Shekerjian, *Uncommon Genius*, (New York: Penguin Books, 1990), 1.

15. L. W. Sumner, *Welfare, Happiness, and Ethics* (Oxford: Clarendon Press, 1996), 145–46.

16. For example, Martin E. P. Seligman, *Authentic Happiness* (New York: Free Press, 2002); and Jonathan Haidt, *The Happiness Hypothesis* (New York: Basic Books, 2006).

17. John Stuart Mill, *Autobiography* (New York: Penguin, 1989 [1873]), 117–18.

18. Gertrude Elion, "Personal Reflections," in *Women in Science and Engineering: Choices for Success*, ed. Cecily Cannan Selby (New York: New York Academy of Sciences, 1999), 18.

19. Alan Gewirth, *Self-Fulfillment* (Princeton, NJ: Princeton University Press, 1998), 14–15, 23, 59.

20. Harry G. Frankfurt, *The Reasons of Love* (Princeton: Princeton University Press, 2004), 6, 14.

21. Mason, *The Value of Creativity*, 235.

22. Mason, *The Value of Creativity*, 235.

23. Mary Midgley, *Wisdom, Information, and Wonder: What Is Knowledge For?* (New York: Routledge, 1989), 4.

24. John Dewey, *Human Nature and Conduct* (New York: Modern Library, 1957 [1922]), 182.

25. John Kekes, *Moral Wisdom and Good Lives* (Ithaca, NY: Cornell University Press, 1995), 9.

26. Robert J. Sternberg, *Wisdom, Intelligence, and Creativity Synthesized* (New York: Cambridge University Press, 2003), 152, 155–56. See also Robert J. Sternberg, "Wisdom and Its Relations to Intelligence and Creativity," in *Wisdom: Its Nature, Origins, and Development*, ed. Robert J. Sternberg (New York: Cambridge University Press, 1990), 142–59.

27. Sternberg, *Wisdom, Intelligence, and Creativity Synthesized*, 178–79.

28. Sternberg, *Wisdom, Intelligence, and Creativity Synthesized*, 180.

29. Sternberg, *Wisdom, Intelligence, and Creativity Synthesized*, 152.

30. Sternberg, *Wisdom, Intelligence, and Creativity Synthesized*, 158.

31. Robert Nozick, *The Examined Life* (New York: Simon and Schuster, 1989), 278.

32. Cf. Levinson, *The Seasons of a Man's Life*, 273.

~

Bibliography

Adams, James L. *Conceptual Blockbusting: A Guide to Better Ideas*. 4th ed. Cambridge, MA: Perseus Publishing, 2001.

Adams, Robert. "Motive Utilitarianism." *Journal of Philosophy* 73 (1976): 467–81.

Adler, Jonathan E. *Belief's Own Ethics*. Cambridge, MA: MIT Press, 2002.

Aguilar, Francis J. *Managing Corporate Ethics*. New York: Oxford University Press, 1994.

Albert, Robert S., ed. *Genius and Eminence*, 2d ed. New York: Pergamon Press, 1992.

Alcoholics Anonymous. New York: Alcoholics Anonymous World Services, 1976.

Allen, Barry. "Forbidding Knowledge." *The Monist* 79, no. 2 (1996): 294–310.

Andre, Judith. "Nagel, Williams, and Moral Luck." Pp. 123–29 in *Moral Luck*, edited by Daniel Statman. Albany: State University of New York Press, 1993.

Andreasen, Nancy C. *The Creating Brain*. New York: Dana Press, 2005.

Arianrhod, Robyn. *Einstein's Heroes: Imagining the World Through the Language of Mathematics*. New York: Oxford University Press, 2005.

Aristotle. *Nicomachean Ethics*, trans. W. D. Ross. Pp. 927–1112 in *Basic Works of Aristotle*, ed. Richard McKeon. New York: Random House, 1941.

Ascheron, Claus, and Angela Kickuth. *Make Your Mark in Science: Creativity, Presenting, Publishing and Patents*. Hoboken, NJ: John Wiley and Sons, 2005.

Atkin, J. Myron, and Paul Black. *Inside Science Education Reform: A History of Curricular and Policy Change*. New York: Teachers College, Columbia University, 2003.

Austin, James H. *Chase, Chance, and Creativity*. Cambridge, MA: MIT Press, 2003.

Auyang, Sunny Y. *Engineering—An Endless Frontier*. Cambridge, MA: Harvard University Press, 2004.

Baer, J. "Domains of Creativity." Pp. 591–96 in *Encyclopedia of Creativity*, ed. Mark A. Runco and Steven R. Pritzker. San Diego: Academic Press, 1999.

Bain, Ken. *What the Best College Teachers Do*. Cambridge, MA: Harvard University Press, 2004.

Barbour, Ian. *Ethics in an Age of Technology*. New York: HarperCollins, 1993.

Barton, Angela Calabrese. *Teaching Science for Social Change*. New York: Teachers College Press, Columbia University, 2003.

Bateson, Mary Catherine. *Composing a Life*. New York: Plume, 1990.

Bernstein, Jeremy. *The Merely Personal: Observations on Science and Scientists*. Chicago: Ivan R. Dee, 2001.

Beveridge, W. I. B. *The Art of Scientific Investigation*. New York: Vintage Books, 1957.

Bird, Alexander. *Thomas Kuhn*. Princeton, NJ: Princeton University Press, 2000.

Bird, Kai, and Martin J. Sherwin. *American Prometheus: The Triumph and Tragedy of J. Robert Oppenheimer*. New York: Alfred A. Knopf, 2005.

Bishop, J. Michael. *How to Win the Nobel Prize*. Cambridge, MA: Harvard University Press, 2003.

Boden, Margaret A. *The Creative Mind*. 2d ed. New York: Routledge, 2004.

Bohm, David. *On Creativity*. New York: Routledge, 1996.

Bohm, David, and F. David Peat. *Science, Order, and Creativity*. 2d ed. New York: Routledge, 2000.

Bok, Derek. *Universities in the Marketplace: The Commercialization of Higher Education*. Princeton, NJ: Princeton University Press, 2003.

Brady, Michael, and Duncan Pritchard, eds. *Moral and Epistemic Virtues*. Malden, MA: Blackwell, 2003.

Broad, William, and Nicholas Wade. *Betrayers of the Truth*. New York: Simon & Schuster, 1982.

Bronowski. *Science and Human Values*. New York: Harper and Row, 1965.

Brown, Laurie M., and John S. Rigden, eds. *"Most of the Good Stuff": Memories of Richard Feynman*, New York: Springer Verlag, 1993.

Bucciarelli, Louis L. *Designing Engineers*. Cambridge, MA: MIT Press, 1994.

Burns, James MacGregor. Forward to *Ethics, the Heart of Leadership*, 2d. ed., edited by Joanne B. Ciulla. Westport, Conn. Praeger, 2004.

———. *Leadership*. New York: Harper and Row, 1978.

Bush, Vannevar. *Science—The Endless Frontier*. Washington, D.C.: NSF, 40th anniversary ed. 1990.

Butler, Samuel. *Erewhon*. New York: New American Library, 1960 [1872].

Bystydzienski, Jill M., and Sharon R. Bird, eds. *Removing Barriers: Women in Academic Science, Technology, Engineering, and Mathematics*. Bloomington: Indiana University press, 2006.

Cajal, Santiago Ramon y. *Advice for a Young Investigator*, trans. Neely Swanson and Larry W. Swanson. Cambridge, MA: MIT Press, 1999.

Card, Claudia. *The Unnatural Lottery: Character and Moral Luck*. Philadelphia: Temple University Press, 1996.

Castel, Boris, and Sergio Sismondo. *The Art of Science*. Ontario, Canada: Broadview Press, 2003.

Chandrasekhar, S. *Truth and Beauty: Aesthetics and Motivations in Science*. Chicago: University of Chicago Press, 1987.
Charles, Daniel. *Master Mind: The Rise and Fall of Fritz Haber, the Nobel Laureate Who Launched the Age of Chemical Warfare*. New York: HarperCollins, 2005.
Chawkins, Steve. "The Baffling Descent of a Nobel Prize Winner. *Los Angeles Times* (August 13, 2005), B1, B7.
Chomsky, Noam. *The Backroom Boys*. London: Fontana/Collins, 1973.
Ciulla, Joanne B. *The Working Life: The Promise and Betrayal of Modern Work*. New York: Times Books, 2000.
Code, Lorraine. *Epistemic Responsibility*. Hanover, NH: University Press of New England, 1987.
Coler, Myron A., ed. *Essays on Creativity in the Sciences*. New York: New York University Press, 1963.
Collins, Jim, and Jerry I. Porras. *Built to Last*. New York: Collins Business, 2002.
Collins, Mary Ann, and Teresa M. Amabile. "Motivation and Creativity." Pp. 297–312 in *Handbook of Creativity*, edited by Robert J. Sternberg. New York: Cambridge University Press, 1999.
Crick, Francis. *What Mad Pursuit: A Personal View of Scientific Discovery*. New York: Basic Books, 1988.
Cropley, A. J. *More Ways than One: Fostering Creativity*. Norwood, NJ: Albex, 1992.
Cropper, William H. *Great Physicists*. New York: Oxford University Press, 2001.
Csikszentmihalyi, Mihaly. *Creativity*. New York: Harper Collins, 1996.
———. *Flow: The Psychology of Optimal Experience*. New York: Harper Perennial, 1990.
Csikszentmihalyi, Mihaly, and R. Keith Sawyer. "Creative Insight: The Social Dimension of a Solitary Moment. Pp. 329–63 in *The Nature of Insight*, edited by R. J. Sternberg and J. E. Davidson. Cambridge, MA: MIT Press, 1995.
Cua, A.S. *Dimensions of Moral Creativity*. University Park: Pennsylvania State University Press, 1978.
Curd, Martin, and J. A. Cover, eds. *Philosophy of Science: The Central Issues*. New York: W.W. Norton and Company, 1998.
Curie, Eve. *Madame Curie*, trans. Vincent Sheean. New York: Doubleday, Doran & Company, 1938.
Darwin, Charles. *The Autobiography of Charles Darwin*, ed. Nora Barlow. New York: W.W. Norton, 1958.
Davis, Michael. "Some Paradoxes of Whistleblowing." *Business and Professional Ethics Journal* 15 (Spring 1996): 3–19.
Dawkins, Richard. *Unweaving the Rainbow: Science, Delusion and the Appetite for Wonder*. Boston: Houghton Mifflin, 1998.
DeGeorge, Richard T. "Ethical Responsibilities of Engineers in Large Organizations." *Business and Professional Ethics Journal* 1 (Fall 1981): 1–14.
Derry, Gregory N. *What Science Is and How It Works*. Princeton, NJ: Princeton University Press, 1999.

DePaul, Michael, and Linda Zagzebski, eds. *Intellectual Virtue*. Oxford: Clarendon Press, 2003.

Desmond, Adrian, and James Moore. *Darwin*. New York: W.W. Norton, 1991.

Dewey, John. *Human Nature and Conduct*. New York: Modern Library, 1957[1922].

Dickenson, Donna. *Risk and Luck in Medical Ethics*. Cambridge: Polity Press, 2003.

Dirac, Paul. "The Evolution of the Physicist's Picture of Nature." *Scientific American*, 208, no. 5 (1963): 45–53.

Djerassi, Carl. *Cantor's Dilemma*. New York: Penguin Books, 1989.

Doris, John M. *Lack of Character: Personality and Moral Behavior*. New York: Cambridge University Press, 2002.

Driscoll, D., and W. M. Hoffman. "From Volumes to Three Words: Texas Instruments." Pp. 623–28 in *Business Ethics*, ed. W. M. Hoffman, R. E. Frederick, and M.S. Schwartz. Boston, MA: McGraw-Hill, 2001.

Duncan, David Ewing. *Masterminds: Genius, DNA, and the Quest to Rewrite Life*. New York: Harper Perennial, 2005.

Einstein, Albert. *The World as I See It*, trans. Alan Harris. New York: Philosophical Library, 1949.

Eldridge, Niles. *Darwin: Discovering the Tree of Life*. New York: W.W. Norton, 2005.

Elion, Gertrude B. "Personal Reflections." Pp. 16–18 in *Women in Science and Engineering: Choices for Success*, edited by Cecily Cannan Selby. New York: New York Academy of Sciences, 1999.

Elliott, Deni, and Judy E. Stern, eds. *Research Ethics*. Hanover, NH: University Press of New England, 1997.

Engler, Gideon. "Aesthetics in Science and in Art." *British Journal of Aesthetics*, 30, no. 1 (1990): 24–33.

Farmelo, Graham, ed. *It Must Be Beautiful: Great Equations of Modern Science*. London: Granta Books, 2003.

———. "Einstein and the Most Beautiful Theories in Physics." *International Studies in the Philosophy of Science* 16, no. 1 (2002): 27–37.

Feist, Gregory J. "The Influence of Personality on Artistic and Scientific Creativity." Pp. 273–296 in *Handbook of Creativity*, ed. Robert J. Sternberg. New York: Cambridge University Press, 1999.

Ferguson, Eugene S. *Engineering and the Mind's Eye*. Cambridge, MA: MIT Press, 1992.

Fesmire, Steven. *John Dewey and Moral Imagination*. Bloomington: Indiana University Press, 2003.

Feyerabend, Paul K. *Against Method: Outline of Anarchistic Theory of Knowledge*. Atlantic Highlands, NJ: Humanities Press, 1975.

Feynman, Richard P. *The Pleasure of Finding Things Out*. Cambridge, MA: Perseus, 1999.

———. *Surely You're Joking, Mr. Feynman!* New York: Bantam, 1986.

Flanagan, Owen. *Varieties of Moral Personality: Ethics and Psychological Realism*. Cambridge, MA: Harvard University Press, 1991.

Florman, Samuel C. *The Existential Pleasures of Engineering*. 2d ed. New York: St. Martin's Griffin, 1994.
Fogler, H. Scott, and Steven E. LeBlanc. *Strategies for Creative Problem Solving*. Upper Saddle River, NJ: Prentice Hall, 1995.
Frankfurt, Harry G. *The Reasons of Love*. Princeton, NJ: Princeton University Press, 2004.
Gabor, Andrea. *Einstein's Wife*. New York: Viking, 1995.
Gardner, Howard. *Creating Minds*. New York: Basic Books, 1993.
———. *Leading Minds*. New York: Basic Books, 1995.
Gardner, John W. *Excellence: Can We Be Equal and Excellent Too?* New York: Harper and Row, 1961.
———. *Self-Renewal*. New York: Harper and Row, 1963.
Gewirth, Alan. *Self-Fulfillment*. Princeton, NJ: Princeton University Press, 1998.
Ghiselin, Brewster, ed. *The Creative Process*. New York: New American Library, 1952.
Gleick, James. *Genius: The Life and Science of Richard Feynman*. New York: Vintage Books, 1992.
Godfrey-Smith, Peter. *Theory and Reality*. Chicago: University of Chicago Press, 2003.
Goethe, Johann Wolfgang Von. *Faust: A Tragedy*, trans. Walter Arndt, edited by Cyrus Hamlin. New York: W.W. Norton, 1975.
Goldberg, Phyllis. "Creeping Toward Inclusivity in Science." Pp. 7–15 in *Women in Science and Engineering: Choices for Success*, edited by Cecily Cannan Selby. New York: New York Academy of Sciences, 1999.
Goldsmith, Barbara. *Obsessive Genius: The Inner World of Marie Curie*. New York: W.W. Norton, 2005.
Goldstein, David. "Richard P. Feynman, Teacher." In *"Most of the Good Stuff": Memories of Richard Feynman*, edited by Laurie M. Brown and John S. Rigden. New York: Springer Verlag, 1993.
Goleman, Daniel. *Emotional Intelligence*. New York: Bantam, 1998.
Goleman, Daniel, Richard Boyatzis, and Annie McKee. *Primal Leadership: Realizing the Power of Emotional Intelligence*. Boston, MA: Harvard Business School Press, 2002.
Goliszek, Andrew. *In the Name of Science*. New York: St. Martin's Press, 2003.
Goodman, Allegra. *Intuition*. New York: Dial Press, 2006.
Graham, Loren R. *The Ghost of the Executed Engineer*. Cambridge, MA: Harvard University Press, 1993.
Greenspan, Nancy Thorndike. *The End of the Certain World: The Life and Science of Max Born*. New York: Basic Books, 2005.
Greenstein, George. *Portraits of Discovery: Profiles in Scientific Genius*. New York: John Wiley and Sons, 1998.
Gribbin, John. *The Scientists*. New York: Random House, 2002.
Haidt, Jonathan. *The Happiness Hypothesis*. New York: Basic Books, 2006.
Hanson, Norwood Russell. *Patterns of Discovery*. Cambridge: Cambridge University Press, 1972.

Hardy, G. H. "A Mathematician's Apology." Pp. 389–94 in *Genius and Eminence*, 2d ed., edited by Robert S. Albert. New York: Pergamon Press, 1992.

Harman, Peter, and Simon Mitton, eds. *Cambridge Scientific Minds*. New York: Cambridge University Press, 2002.

Hassard, Jack. *The Art of Teaching Science*. New York: Oxford University Press, 2005.

Herken, Gregg. *Brotherhood of the Bomb: The Tangled Lives and Loyalties of Robert Oppenheimer, Ernest Lawrence, and Edward Teller*. New York: Henry Holt and Company, 2002.

Hickman, Larry A. *John Dewey's Pragmatic Technology*. Bloomington: Indiana University Press, 1990.

Hoddeson, Lillian, and Vicki Daitch. *True Genius: The Life and Science of John Bardeen*. Washington, DC: Joseph Henry Press, 2002.

Hodge, Jonathan, and Gregory Radick, eds. *The Cambridge Companion to Darwin*. New York: Cambridge University Press, 2003.

Homer-Dixon, Thomas. *The Ingenuity Gap*. New York: Vintage Books, 2002.

Hooker, Charles, and Mihaly Csikszentmihalyi. "Flow, Creativity, and Shared Leadership." Pp. 217–34 in *Shared Leadership: Reframing the Hows and Whys of Leadership*, ed. Craig L. Pearce and Jay A. Conger. Thousand Oaks, CA: Sage Publications, 2003.

Howe, Michael J. A. *Genius Explained*. New York: Cambridge University Press, 1999.

Huizenga, John R. *Cold Fusion: The Scientific Fiasco of the Century*. New York: Oxford University Press, 1993.

Hull, David L. *Science as a Process*. Chicago: University of Chicago Press, 1988.

Hutcheson, Francis. *An Inquiry Concerning Beauty, Order, Harmony, Design*, edited by Peter Kivy. The Hague: Martinus Nijhoff, 1973.

Irwin, Aisling. "An Environmental Fairy Tale: The Molina-Rowland Chemical Equations and the CFC Problem." Pp. 87–109 in *It Must Be Beautiful: Great Equations of Modern Science*, edited by Graham Farmelo. London: Granta Books, 2003.

Israel, Paul. *Edison: A Life of Invention*. New York: John Wiley, 1998.

James, Gene G. "Whistle Blowing: Its Moral Justification." Pp. 291–302 in *Business Ethics*, 4th ed., edited by W. Michael Hoffman, Robert E. Frederick, and Mark S. Schwartz. Boston: McGraw Hill, 2001.

Jamison, Kay Redfield. *Exuberance: The Passion for Life*. New York: Alfred A. Knopf, 2004.

Janis, Irving L. "Groupthink." *Psychology Today* (November 1971): 271–79.

Janovy, John, Jr. *On Becoming a Biologist*. 2d ed. Lincoln: University of Nebraska Press, 2004.

John-Steiner, Vera. *Creative Collaboration*. New York: Oxford University Press, 2000.

Johnson, Deborah G. "Forbidden Knowledge and Science as Professional Activity." *Monist* 79, no. 2 (1996): 197–217.

Jones, James H. *Bad Blood: The Tuskegee Syphilis Experiment*. New York: Free Press, 1993.

Judson, Horace Freeland. *The Great Betrayal: Fraud in Science*. Orlando, FL: Harcourt, 2004.

Kant, Immanuel. *Practical Philosophy*, trans. and ed. Mary J. Gregor. New York: Cambridge University Press, 1996.
Katz, Ralph, ed. *Managing Professionals in Innovative Organizations*, Cambridge, MA: Ballinger Publishing, 1988.
Kekes, John. *The Art of Life*. Ithaca, NY: Cornell University Press, 2002.
———. *Moral Wisdom and Good Lives*. Ithaca, NY: Cornell University Press, 1995.
Keller, Evelyn Fox. *A Feeling for the Organism: The Life and Work of Barbara McClintock*. New York: W. H. Freeman, 1983.
Kellerman, Barbara. *Bad Leadership*. Boston, MA: Harvard Business School Press, 2004.
Kettlewell, Caroline. *Electric Dreams*. New York: Carroll and Graf Publishers, 2004.
Kidder, Tracy. *The Soul of a New Machine*. New York: Avon, 1981.
Kitcher, Philip. *Science, Truth, and Democracy*. New York: Oxford University Press, 2001.
Klemke, E. D., Robert Hollinger, and David Wyss Rudge, eds. *Introductory Readings in the Philosophy of Science*. 3d ed. Amherst, NY: Prometheus Books, 1998.
Kluger, Jeffrey. *Splendid Solution: Jonas Salk and the Conquest of Polio*. New York: Berkley Books, 2004.
Koertge, Noretta, ed. *Scientific Values and Civic Virtues*. New York: Oxford University Press, 2005.
Koestler, Arthur. *The Act of Creation*. New York: Dell, 1963.
Kopp, Wendy. *One Day, All Children*. New York: Public Affairs, 2003.
Koppang, Haavard, and Mike W. Martin. "On Moralizing in Business Ethics." *Business and Professional Ethics Journal*, 23, no. 3 (2004): 107–14.
Kuhn, Thomas S. *The Road Since Structure: Philosophical Essays, 1970– 1993*, edited by James Conant and John Haugeland. Chicago: University of Chicago Press, 2000.
———. *The Structure of Scientific Revolutions*. 3d ed. Chicago: University of Chicago Press, 1996.
Lamb, David. *Discovery, Creativity and Problem-Solving*. Brookfield: Avebury, 1991.
Lang, Berel. *Act and Idea in the Nazi Genocide*. Chicago: University of Chicago Press, 1990.
Lax, Eric. *The Mold in Dr. Florey's Coat*. New York: Henry Holt and Company, 2005.
Levinson, Daniel J. *The Seasons of a Man's Life*. New York: Ballantine Books, 1978.
Levi-Strauss, Claude. *The Savage Mind*. Chicago: University of Chicago Press, 1966.
Levitt, Norman. *Prometheus Bedeviled: Science and the Contradictions of Contemporary Culture*. New Brunswick, NJ: Rutgers University Press, 1999.
Lewis, E. E. *Masterworks of Technology: The Story of Creative Engineering, Architecture, and Design*. Amherst, NY: Prometheus Books, 2004.
Lifton, Robert Jay. *The Nazi Doctors: Medical Killing and the Psychology of Genocide*. New York: Basic Books, 1986.
Lightman, Alan. *A Sense of the Mysterious: Science and the Human Spirit*. New York: Pantheon Books, 2005.
Longino, Helen E. *The Fate of Knowledge*. Princeton, NJ: Princeton University Press, 2002.

Lynch, Michael P., ed. *The Nature of Truth*. Cambridge, MA: MIT Press, 2001.
Macfarlane, Gwyn. *Alexander Fleming*. Cambridge, MA: Harvard University Press, 1984.
MacIntyre, Alasdair. *After Virtue*. 2d ed. Notre Dame, IN: University of Notre Dame Press, 1984.
Maddox, Brenda. *Rosalind Franklin: The Dark Lady of DNA*. New York: Perennial, 2002.
Martin, Mike W. *Meaningful Work: Rethinking Professional Ethics*. New York: Oxford University Press, 2000.
———. "Moral Creativity in Science and Engineering." *Science and Engineering Ethics* 12, no. 3 (2006): 421–33.
———. "Personal Meaning and Ethics in Engineering." *Science and Engineering Ethics* 8 (2002): 545–60.
———. "Paradoxes of Moral Motivation." *Journal of Value Inquiry*. Forthcoming.
———. *Self-Deception and Morality*. Lawrence: University Press of Kansas, 1986.
———. *Virtuous Giving: Philanthropy, Voluntary Service, and Caring*. Bloomington: Indiana University Press, 1994.
Martin, Mike W., and Roland Schinzinger. *Ethics in Engineering*. 4th ed. Boston: McGraw-Hill, 2005.
Mason, John Hope. *The Value of Creativity: The Origins and Emergence of a Modern Belief*. Burlington, VT: Ashgate, 2003.
Matthews, Jay. *Escalante*. New York: Henry Holt, 1988.
McAllister, James W. *Beauty and Revolution in Science*. Ithaca, NY: Cornell University Press, 1996.
———. "Introduction to Recent Work on Aesthetics of Science." *International Studies in the Philosophy of Science* 16, no. 1 (2002): 7–11.
McElheny, Victor K. *Watson and DNA: Making a Scientific Revolution*. New York: Basic Books, 2004.
McGrayne, Sharon Bertsch. *Nobel Prize Women in Science*. 2d ed. Washington, DC: Joseph Henry Press, 1998.
———. *Prometheans in the Lab: Chemistry and the Making of the Modern World*. New York: McGraw-Hill, 2001.
Merton, Robert K. "Making It Scientifically." Pp. 213–18 in *The Double Helix*, ed. Gunther S. Stent. New York: W.W. Norton, 1980.
———. "Priorities in Scientific Discovery." Pp. 286–324 in Robert K. Merton, *The Sociology of Science*, ed. Norman W. Storer. Chicago: University of Chicago Press, 1973.
———. *Social Theory and Social Structure*. New York: Free Press, 1968.
Merton, Robert K., and Elinor Barber. *The Travels and Adventures of Serendipity*. Princeton, NJ: Princeton University Press, 2004.
Midgley, Mary. "Creation and Originality." Pp. 43–58 in Mary Midgely, *Heart and Mind: The Varieties of Moral Experience*. New York: St. Martin's Press, 1981.
———. *Wisdom, Information, and Wonder: What Is Knowledge For?* New York: Routledge, 1989.

Mill, John Stuart. *Autobiography*. New York: Penguin Books, 1989.
———. *On Liberty*. Indianapolis, IN: Hackett Publishing, 1978.
Miller, Arthur I. *Insights of Genius: Imagery and Creativity in Science and Art*. Cambridge, MA: MIT Press, 2000.
Monsma, Stephen, C. Christians, E.R. Dykema, L. Arie, S. Egbert, and V.P. Lambert. *Responsible Technology: A Christian Perspective*. Grand Rapids, MI: William B. Eerdmans Publishing, 1986.
Mooney, Chris. *The Republican War on Science*. New York: Basic Books, 2005.
Moriarty, Gene. "Ethics, *Ethos* and the Professions: Some Lessons from Engineering. *Professional Ethics* 4 (1995): 75–93.
Moulthrop, Daniel, Ninive Clements Calegari, and Dave Eggers. *Teachers Have It Easy: The Big Sacrifices and Small Salaries of America's Teachers*. New York: The New Press, 2005.
Munson, Ronald. *Intervention and Reflection*. 7th ed. Belmont, CA: Wadsworth, 2004.
Nagel, Thomas. *Mortal Questions*. New York: Cambridge University Press, 1979.
Nasar, Sylvia. *A Beautiful Mind*. New York: Simon & Schuster, 1998.
National Academy of Engineering. http://www.greatachievements.org.
Newbold, Heather. *Life Stories: World-Renowned Scientists Reflect on Their Lives and the Future of Life on Earth*. Berkeley: University of California Press, 2000.
Newton-Smith, W. H., ed. *A Companion to the Philosophy of Science*. Malden, MA: Blackwell, 2000.
Nickerson, Raymond S. "Enhancing Creativity." Pp. 392–430 in *Handbook of Creativity*, edited by Robert J. Sternberg. New York: Cambridge University Press, 1999.
Nozick, Robert. *The Examined Life*. New York: Simon & Schuster, 1989.
Pacey, Arnold. *Meaning in Technology*. Cambridge, MA: MIT Press, 1999.
Pais, Abraham. *Subtle Is the Lord: The Science and Life of Albert Einstein*. Oxford: Oxford University Press, 1982.
Parsons, Keith, ed. *The Science Wars*. Amherst, NY: Prometheus Books, 2003.
Pearson, Greg, and A. Thomas Young, eds. *Technically Speaking: Why All Americans Need to Know More about Technology*. Washington, DC: National Academy Press, 2002.
Penslar, Robin Levin, ed. *Research Ethics: Cases and Material*. Bloomington: Indiana University Press, 1995.
Perkins, David N. *The Eureka Effect: The Art and Logic of Breakthrough Thinking*. New York: W.W. Norton, 2000.
———. *The Mind's Best Work*. Cambridge, MA: Harvard University Press, 1981.
Peters, Thomas J. "A Skunkworks Tale." Pp. 433–41 in *Managing Professionals in Innovative Organizations*, edited by Ralph Katz. Cambridge, MA: Ballinger Publishing Company, 1988.
Peters, Thomas J., and Robert H. Waterman. *In Search of Excellence Lessons from America's Best-Run Companies*. New York: Warner Books, 1982.
Pfenninger, Karl H., and Valerie R. Shubik, eds. *The Origins of Creativity*. New York: Oxford University Press, 2001.

Pickstone, John V. *Ways of Knowing: A New History of Science, Technology and Medicine*. Chicago: University of Chicago Press, 2000.
Pincoffs, Edmund L. *Quandaries and Virtues*. Lawrence: University Press of Kansas, 1986.
Polanyi, Michael. *Personal Knowledge*. Chicago: University of Chicago Press, 1958.
Popper, Karl R. *The Logic of Scientific Discovery*. New York: Harper and Row, 1965.
Price, Terry L. *Understanding Ethical Failures and Leadership*. New York: Cambridge University Press, 2006.
Quinn, Susan. *Marie Curie: A Life*. Reading, MA: Perseus, 1995.
Rachels, James. *The Elements of Moral Philosophy*. 4th ed. New York: McGraw-Hill, 2003.
Regis, Ed. *Who Got Einstein's Office?: Eccentricity and Genius at the Institute for Advanced Study*. Reading, MA: Addison-Wesley, 1987.
Rescher, Nicholas. "Forbidden Knowledge: Moral Limits of Scientific Research." Pp. 1–16 in Nicholas Rescher, *Forbidden Knowledge and Other Essays on the Philosophy of Cognition*. Dordrecht: D. Reidel Publishing, 1987.
———. *Luck*. Pittsburgh: University of Pittsburgh Press, 1995.
Resnik, David B. *The Ethics of Science*. New York: Routledge, 1998.
Richardson, W. Mark, and Gordy Slack, eds. *Faith in Science: Scientists Search for Truth*. New York: Routledge, 2001.
Ridley, Matt. *Francis Crick: Discoverer of the Genetic Code*. New York: HarperCollins, 2006.
Roberts, Robert C., and W. Jay Wood. "Humility and Epistemic Goods." Pp. 257–79 in *Intellectual Virtue*, ed. Michael DePaul and Linda Zagzebski. Oxford: Clarendon Press, 2003.
Roberts, Royston M. *Serendipity: Accidental Discoveries in Science*. New York: John Wiley and Sons, 1989.
Rogers, Carl. "Toward a Theory of Creativity." Pp. 296–305 in *The Creativity Question*, edited by Albert Rothenberg and Carl R. Hausman. Durham, NC: Duke University Press, 1976.
Rollin, Bernard E. *Science and Ethics*. New York: Cambridge University Press, 2006.
Rosenberg, Alex. *Philosophy of Science*. 2d ed. New York: Routledge, 2005.
Rosner, Lisa, ed. *The Technological Fix: How People Use Technology to Create and Solve Problems*. New York: Routledge, 2004.
Rost, Joseph C. *Leadership for the Twenty-First Century*. Westport, Conn.: Praeger, 1991.
Roth, Wolff-Michael, and Angela Calabrese Barton. *Rethinking Scientific Literacy*. New York: Routledge Falmer, 2004.
Rothenberg, Albert, and Carl R. Hausman, eds. *The Creativity Question*. Durham, NC: Duke University Press, 1976.
Rowland, Sherwood. "Ozone Hole." Pp 134–40 in *Life Stories: World-Renowned Scientists Reflect on Their Lives and the Future of Life on Earth*, edited by Heather Newbold. Berkeley: University of California Press, 2000.

Runco, Mark A., and Steven R. Pritzker, eds. *Encyclopedia of Creativity*. San Diego: Academic Press, 1999.

Runco, Mark A., and Ruth Richards. *Eminent Creativity, Everyday Creativity, and Health*. Greenwich, CT: Ablex Publishing Corporation, 1997.

Russell, Bertrand. *The Conquest of Happiness*. New York: Liveright, 1930.

Sagan, Carl. *The Demon-Haunted World: Science as a Candle in the Dark*. New York: Ballantine Books, 1996.

Salmon, M. H., J. Earman, C. Glymour, J. G. Lennox, P. Machamer, J. E. McGuire, J. D. Norton, W. C. Salmon, and K. F. Schaffner, eds. *Introduction to the Philosophy of Science*. Indianapolis, IN: Hackett Publishing, 1992.

Sapienza, Alice M. *Managing Scientists: Leadership Strategies in Scientific Research*. 2d ed. New York: Wiley-Liss, 2004.

Sartre, Jean-Paul. *Being and Nothingness*, trans. Hazel E. Barnes. New York: Washington Square Press, 1966.

———. "Existentialism Is a Humanism," trans. P. Mairet. Pp. 345–69 in *Existentialism from Dostoevsky to Sartre*, ed. Walter Kaufmann. New York: New American Library, 1975.

Sawyer, R. Keith. *Explaining Creativity: The Science of Human Innovation*. New York: Oxford University Press, 2006.

Scheffler, Israel. *The Anatomy of Inquiry: Philosophical Studies in the Theory of Science*. Indianapolis, IN: Bobbs-Merrill, 1963.

Schiebinger, Londa. *Has Feminism Changed Science?* Cambridge, MA: Harvard University Press, 1999.

Schon, Donald A. *The Reflective Practitioner: How Professionals Think in Action*. New York: BasicBooks, 1983.

Schweitzer, Albert. *Goethe: Five Studies*, trans. Charles R. Joy. Boston: Beacon Press, 1961.

Seebauer, Edmund G., and Robert L. Barry. *Fundamentals of Ethics for Scientists and Engineers*. New York: Oxford University Press, 2001.

Selby, Cecily Cannan, ed. *Women in Science and Engineering: Choices for Success*. New York: New York Academy of Sciences, 1999.

Seligman, Martin E. P. *Authentic Happiness*. New York: Free Press, 2002.

Sennett, Richard. *The Corrosion of Character: The Personal Consequences of Work in the New Capitalism*. New York: W.W. Norton & Company, 1998.

Shapero, Albert. "Managing Creative Professionals." Pp. 215–22 in *Managing Professionals in Innovative Organizations*, ed. Ralph Katz. Cambridge, MA: Ballinger Publishing, 1988.

Shapiro, Gilbert. *A Skeleton in the Darkroom: Stories of Serendipity in Science*. San Francisco: Harper & Row, 1986.

Shapiro, Harold T. *A Larger Sense of Purpose: Higher Education and Society*. Princeton, NJ: Princeton University Press, 2005.

Shattuck, Roger. *Forbidden Knowledge*. San Diego: Harcourt Brace & Company, 1996.

Shekerjian, D. *Uncommon Genius*. New York: Penguin Books, 1990.

Shrader-Frechette, Kristin. *Ethics of Scientific Research*. Lanham, MD.: Rowman & Littlefield, 1994.

Sidgwick, Henry. *The Methods of Ethics*. 7th ed. Chicago: University of Chicago Press, 1962.

Simon, Bart. *Undead Science: Science Studies and the Afterlife of Cold Fusion*. New Brunswick, NJ: Rutgers University Press, 2002.

Simon, Stephanie. "Their Own Version of a Big Bang." *Los Angeles Times* (February 1, 2006): 1, 16.

Simonton, Dean Keith. *Creativity in Science*. New York: Cambridge University Press, 2004.

———. *Origins of Genius: Darwinian Perspectives on Creativity*. New York: Oxford University Press, 1999.

Singer, Peter. "Ethics and the Limits of Scientific Freedom." *Monist* 79, no. 2 (1996): 218–29.

Singer, S. Jonathan. *The Splendid Feast of Reason*. Berkeley: University of California Press, 2001.

Sinsheimer, Robert L. "Review of *The Double Helix*." Pp. 191–94 in *The Double Helix*, ed. Gunther S. Stent. New York: W.W. Norton, 1980.

Slote, Michael. *Goods and Virtues*. Oxford: Clarendon Press, 1983.

Smith, Adam. *An Inquiry into the Nature and Causes of the Wealth of Nations*. New York: Oxford University Press, 1976.

Smith, J. S. *Patenting the Sun: Polio and the Salk Vaccine*. New York: William Morrow and Company, 1990.

Stangroom, Jeremy, ed. *What Scientists Think*. New York: Routledge, 2005.

Statman, Daniel. "Introduction." Pp. 1–34 in *Moral Luck*, ed. Daniel Statman. Albany: State University of New York Press, 1993.

Steiner, G.A. *The Creative Organization*. Chicago: University of Chicago Press, 1965.

Stent, Gunther S. "Meaning in Art and Science." Pp. 31–42 in *The Origins of Creativity*, ed. Karl H. Pfenninger and Valerie R. Shubik. New York: Oxford University Press, 2001.

Sternberg, Robert J., ed. *Handbook of Creativity*. New York: Cambridge University Press, 1999.

———, ed. *The Nature of Creativity*. New York: Cambridge University Press, 1988.

———. "Wisdom and Its Relations to Intelligence and Creativity." Pp. 142–59 in *Wisdom: Its Nature, Origins, and Development*, edited by Robert J. Sternberg. New York: Cambridge University Press, 1990.

———. *Wisdom, Intelligence, and Creativity Synthesized*. New York: Cambridge University Press, 2003.

Sternberg, R. J., and J. E. Davidson, eds. *The Nature of Insight*. Cambridge, MA: MIT Press, 1995.

Sternberg, Robert J., and Jennifer Jordan, eds. *A Handbook of Wisdom: Psychological Perspectives*. New York: Cambridge University Press, 2005.

Steup, Matthias, ed. *Knowledge, Truth, and Duty: Essays on Epistemic Justification, Responsibility, and Virtue*. New York: Oxford University Press, 2001.

Stevenson, Robert Louis. *Dr. Jekyll and Mr. Hyde*. New York: Bantam, 1981.
Stout, Jeffrey. *Ethics After Babel: The Languages of Morals and Their Discontents*, rev. ed. Princeton, NJ: Princeton University Press, 2001.
Sumner, L. W. *Welfare, Happiness, and Ethics*. Oxford: Clarendon Press, 1996.
Taubes, Gary. *Bad Science: The Short Life and Weird Times of Cold Fusion*. New York: Random House, 1993.
Taylor, Charles. *The Ethics of Authenticity*. Cambridge, MA: Harvard University Press, 1992.
Thoreau, Henry David. *Walden and Civil Disobedience*. New York: Penguin, 1983.
Tiles, Mary, and Hans Oberdiek. *Living in a Technological Culture*. London: Routledge, 1995.
Tobin, Kenneth, Rowhea Elmesky, and Gale Seiler, eds. *Improving Urban Science Education*. Lanham. MD: Rowman and Littlefield, 2005.
Tomas, Vincent. "Introduction." Pp. 1–3 in *Creativity in the Arts*, edited by Vincent Tomas. Englewood Cliffs, NJ: Prentice-Hall, 1964.
Trilling, Lionel. *Sincerity and Authenticity*. Cambridge, MA: Harvard University Press, 1971.
Vincenti, Walter G. *What Engineers Know and How They Know It*. Baltimore, MD: Johns Hopkins University Press, 1990.
Wadman, Meredith. "One in Three Scientists Confesses to Having Sinned." *Nature* 435 (June 9, 2004): 718–19.
Walker, Margaret Urban. *Moral Contexts*. Lanham, MD: Rowman & Littlefield, 2003.
Wallace, James D. *Moral Relevance and Moral Conflict*. Ithaca, NY: Cornell University press, 1988.
Waller, J. *Einstein's Luck: The Truth Behind Some of the Greatest Scientific Discoveries*. New York: Oxford University Press, 2002.
Walpole, Horace. Letter to Horace Mann, January 28, 1954. Pp. 407–11 in *The Yale Edition of Horace Walpole's Correspondence*, edited by W. S. Lewis. New Haven, CT: Yale University Press, 1937–1983, vol. 20.
Warnock, Mary. *Nature and Morality: Recollections of a Philosopher in Public Life*. New York: Continuum, 2003.
Washburn, Jennifer. *University, Inc.: The Corporate Corruption of American Higher Education*. New York: Basic Books, 2005.
Watson, James D. *The Double Helix*. Pp. 3–133 in *The Double Helix*, ed. Gunther S. Stent. New York: W.W. Norton, 1980.
Weber, Max. "Science as a Vocation." Pp. 1–31 in *The Vocation Lectures*, trans. Rodney Livingstone, edited by David Owen and Tracy B. Strong. Indianapolis, IN: Hackett Publishing 2004.
Weinberg, Alvin M. "Can Technology Replace Social Engineering?" *University of Chicago Magazine* 59 (October 1966): 6–10.
Weisberg, Robert W. *Creativity: Genius and Other Myths*. New York: W.H. Freeman and Company, 1986.
Welch, Jack, with John A. Byrne. *Jack: Straight From the Gut*. New York: Warner Books, 2001.

Wells, H. G. *Meanwhile*. London: Ernest Benn Limited, 1927.
Whitbeck, Caroline. *Ethics in Engineering Practice and Research*. New York: Cambridge University Press, 1998.
Williams, Bernard. *Ethics and the Limits of Philosophy*. Cambridge, MA: Harvard University Press, 1985.
———. *Moral Luck*. New York: Cambridge University Press, 1993.
———. "Persons, Character and Morality." Pp. 1–19 in Bernard Williams, *Moral Luck*. New York: Cambridge University Press, 1981.
Williams, Wendy M., and Lana T. Yang. "Organizational Creativity." Pp. 373–91 in *Handbook of Creativity*, edited by Robert J. Sternberg. New York: Cambridge University Press, 1999.
Wilson, Edward O. *Consilience: The Unity of Knowledge*. New York: Vintage Books, 1996.
Wolf, Susan. "Happiness and Meaning: Two Aspects of the Good Life." Pp. 207–25 in *Self-Interest*, edited by Ellen Frankel Paul, Fred D. Miller, Jr., and Jeffrey Paul. New York: Cambridge University Press, 1997.
Wolpert, Lewis, and Alison Richards. *Passionate Minds: The Inner World of Scientists*. New York: Oxford University Press, 1997.
Young-Bruehl, Elisabeth. *Creative Characters*. New York: Routledge, 1991.
Zagzebski, Linda Trinkaus. *Virtues of the Mind*. New York: Cambridge University Press, 1996.
Zygmont, Jeffrey. *Microchip: An Idea, Its Genesis, and the Revolution It Created*. New York: Perseus Publishing, 2003.

Index

About the Author

Mike W. Martin is professor of philosophy at Chapman University (Orange, California). He is the author of seventy essays, most in applied ethics; and he is the author, coauthor, or editor of eleven books, including *From Morality to Mental Health: Virtue and Vice in a Therapeutic Culture* (Oxford University Press, 2006), *Ethics in Engineering*, 4th edition (with Roland Schinzinger, McGraw-Hill, 2005), *Meaningful Work: Rethinking Professional Ethics* (Oxford University Press, 2000), and *Albert Schweitzer's Reverence for Life: Ethical Idealism and Self-Realization* (Ashgate Publishing, forthcoming).